Building the homestead

African Studies Centre
Research Series
16/2001

Building the homestead

Agriculture, labour and beer in South Africa's Transkei

Patrick McAllister

Routledge
Taylor & Francis Group
LONDON AND NEW YORK

First published 2001 by Ashgate Publishing

Reissued 2018 by Routledge
2 Park Square, Milton Park, Abingdon, Oxon OX14 4RN
711 Third Avenue, New York, NY 10017, USA

Routledge is an imprint of the Taylor & Francis Group, an informa business

Cover photo: *A small work party removing weeds and grass from a garden in Shixini* (photo: Patrick McAllister)

A Library of Congress record exists under LC control number: 2002391676

ISBN 13: 978-1-138-70362-9 (hbk)
ISBN 13: 978-1-315-20306-5 (ebk)

Contents

Maps, Tables and Photographs

Preface

South Africa has generated enormous research interest over the past ten years but its rural areas have attracted less attention. Professor McAllister's work provides striking evidence of the variation and unevenness of rural livelihood in South Africa, demonstrating that agricultural activities directed at subsistence, notably maize production, continue to thrive in Shixini, and may even agronomically harbour the potential for extended commodity production.

This study builds on research findings emanating from the De-agrarianization and Rural Employment (DARE) research programme co-ordinated by the African Studies Centre, Leiden, and financed by the Netherlands' Ministry of Foreign Affairs. The DARE programme involved four core research teams in Ethiopia, Nigeria, Tanzania and South Africa. Professor McAllister's study was linked to the work of his fellow DARE research team members, Leslie Bank and Wele Manona, based at the Institute of Social and Economic Research at Rhodes University, South Africa.

The DARE programme's research objectives have been to compare and contrast the process of de-agrarianization in various rural areas of Africa in terms of economic activity reorientation, occupational adjustment, social identification, and spatial relocation of rural dwellers away from strictly peasant modes of livelihood. The extent and nature of non-agricultural activities in rural areas have been examined in relation to the influence of risk on rural household production and exchange. As has been amply demonstrated in the case of Shixini, and mirrored in all the other DARE case-study findings, de-agrarianization processes in Africa have in fact reaffirmed rather than undermined the importance of the 'subsistence fallback', namely, rural dwellers' reliance on own-farm produced staple foodstuffs.

The overall findings from the DARE programme are intended to provide insight into the processes of change which are shaping the livelihood prospects of African rural and urban dwellers in the twenty-first century. It is hoped that the knowledge gained may be useful for formulating more effective developmental policies to assist in short-circuiting Sub-Saharan Africa's current economic and political vulnerabilities.

Deborah Fahy Bryceson
DARE Research Programme Coordinator
African Studies Centre, Leiden

Acknowledgements

This book is an outcome of my involvement during the period between 1995 and 1999 in the De-agrarianization and Rural Employment (DARE) research programme co-ordinated from the African Studies Centre in Leiden. I am grateful to Deborah Bryceson and other members of the research team for the opportunity to return to some of the agrarian issues that have interested me for a number of years, although my findings contrast rather markedly with those of the other areas included in the project (see Bryceson 1998). I am also grateful to the African Studies Centre for a Research Fellowship in 1999, which enabled me to start writing up the material on which this volume is based. This material was obtained partly during two short fieldwork periods in 1996 and 1998, during which Mr. Cecil Nonqane of the Albany Museum, Grahamstown, assisted me. But I have drawn also on earlier material collected at various points in my association with the people of Shixini in the Gatyana (Willowvale) district of South Africa's Eastern Cape Province, in what was the Transkei 'homeland',[1] which commenced in 1976, and on which many of my colleagues have commented from time to time. I would like to mention, in particular, those with whom I worked in the Department of Anthropology at Rhodes University, Grahamstown, during the period 1981-1990, Chris de Wet, Robin Palmer and Michael Whisson. I have also drawn on data collected

[1] Transkei was formerly one of the nominally independent 'homelands' established in terms of apartheid ideology. It is now part of the Eastern Cape Province, but as a distinct part of the Eastern Cape is still often referred to by its older name, or by the more tedious 'former Transkei', which I avoid. English and Xhosa names for districts have co-existed for many years, with the Xhosa names now enjoying official status. In this book I refer to the district as Gatyana and the town within it as 'Willowvale town'.

in Shixini by two post-graduate students, Maura Andrew and Gavin Heron, who completed Master of Arts dissertations in 1990 and 1992, on land use patterns and co-operative work, respectively. I am grateful to Ben Cousins for assistance in identifying material relevant to the policy and development implications of the findings presented below. In Shixini itself my hosts, Nothusile Tshemese, her late husband, Mzilikazi, as well as their kin and neighbours, have for long been unfailingly helpful, generous and tolerant of my intrusions.

During my association with the people of Shixini my approach has been the conventional anthropological one of participant-observation and informal interview, though from time to time I have conducted household censuses or surveys of various kinds. My visits to Shixini have concentrated on the three southeastern and adjacent sub-wards of Nompha, Ndlelibanzi and Folokhwe, though it is in the nature of this kind of fieldwork for data to be drawn from many people and events outside of these three areas as well as from those within them. For the purposes of the DARE study a basic census was done in 1998 of all the 194 homesteads in these three sub-wards to provide information that could be compared with census-type surveys done in 1990 (in the same three sub-wards) and in1976 (in Folokhwe sub-ward). I am grateful to a Folokhwe resident and friend, Masilingane Mhlakaza, who ably assisted me with the 1998 census. In what follows, some reference is made to findings derived from census figures, but most of the data is of a qualitative rather than a quantitative kind

Finally, I want to express my gratitude to Mieke Zwart (African Studies Centre) who took care of the layout work, and to Nel de Vink who drew the maps.

1

Introduction

Fieldwork in Shixini

When I started fieldwork in Shixini, an Administrative Area in Willowvale/Gatyana district in the Eastern Cape Province (see Map 1), in1976, I soon found that I had little option but to start attending beer parties, because this is where most of the Xhosa people whom I hoped to speak to seemed to spend a lot of their time. These parties were held in individual homesteads, and at first I imagined that they were similar to you or I hosting a social function of some sort; in other words, that they were unremarkable. After all, African ethnographies are littered with references to beer drinking, but mostly without any detailed description of beer-drinking events or analysis of their social significance. I thought that perhaps this was why there was hardly anything substantial about beer drinking in the writings of the South African ethnographers whose work had read as an undergraduate, and revisited in preparation for fieldwork.[1]

[1] There were some exceptions to this. For example, the Kriges had quite a lot to say about beer drinking among the Lovedu, and Davies had published an early article about Bomvana beer drinks (Krige and Krige 1943; Davies 1927).

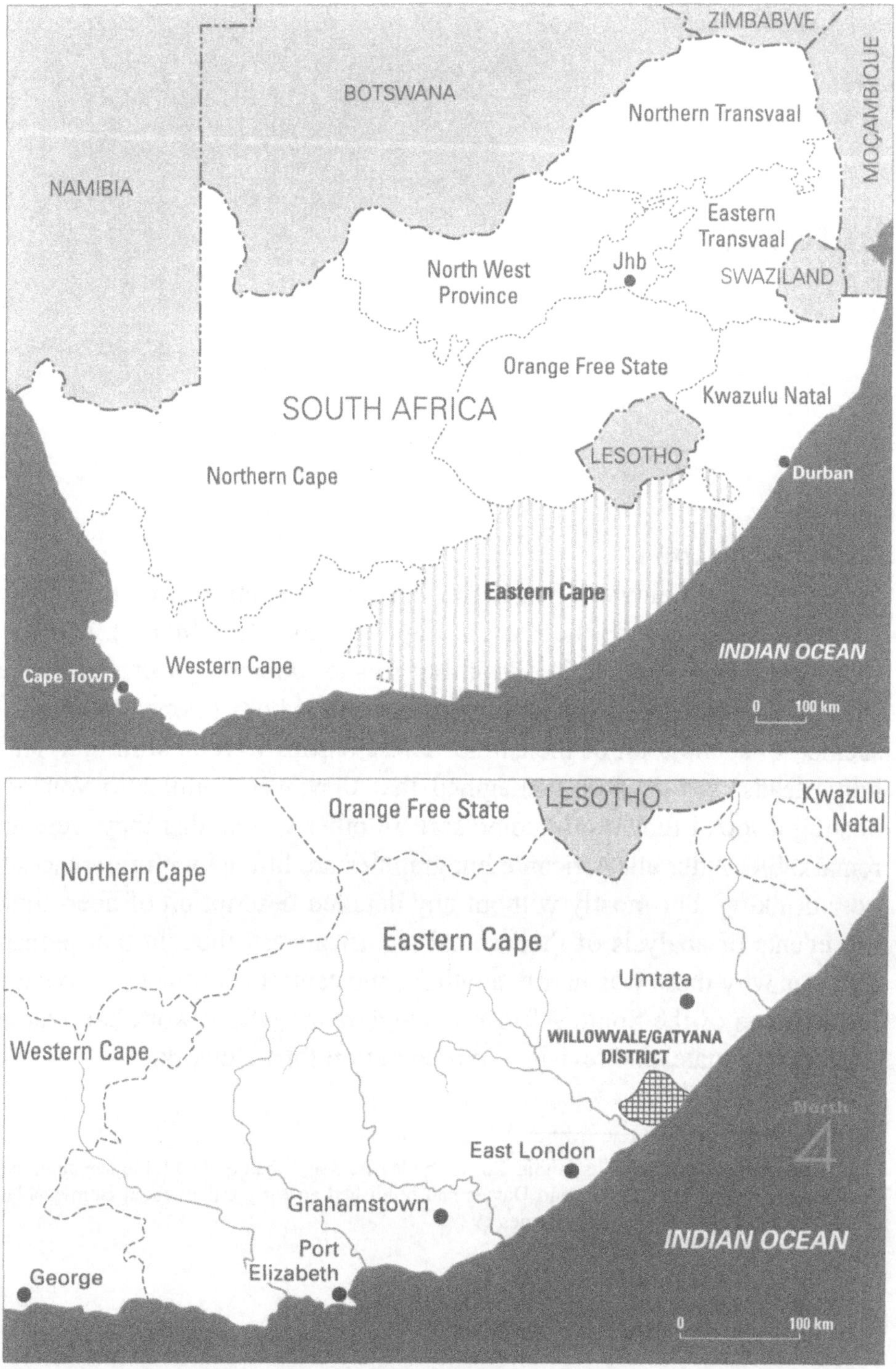

Map 1 Eastern Cape Province, South Africa

However, after a month or two I noticed that the beer drinks I was somewhat reluctantly attending were socially important, patterned in interesting and socially significant ways, and worth paying attention to.

For one thing, neighbours were not specifically invited to beer drinks but they always turned up anyway, and they acted as if it was their party, not that of the head of the host homestead. Soon after arriving, they would demand something to drink, rather belligerently, and the host always complied. They would then decide that it was time that they took control of the proceedings (and the beer) themselves, and that the host could not be left in charge. They took on the role of allocating the beer, deciding who got beakers of beer to drink, in what order and in what quantity. They even told the guests where they were and were not allowed to sit. Some people were allowed into the hut where the beer was kept under the watchful eye of the neighbours who had taken charge; others, usually including all women, had to sit outside, each person in his or her group, with each group allocated to a specific seating place. Those who took charge clearly regarded the host as not socially competent enough to be given these important tasks, and if he tried to intervene, they would tell him, more or less politely, to sit down and be quiet; and he would do so without too much protest. It was always a he, even if the head of the homestead was a woman, who would be represented by a stand-in for her absent or deceased spouse.[2]

This group of neighbours then proceeded to debate and argue fiercely among themselves about who should get to drink what was on offer, and in what order. Long arguments, intense and sometimes heated, conducted through skilful oratory spiced with humour, took place between the local and visiting groups too, and revolved around the places and the beer that the visitors were allocated. In the meantime the self-appointed hosts made sure that they themselves had plenty of beer, and this helped to prolong the argument and add to the din. They lorded it over the homestead head, and over his friends and his kin, and at some point they invariably insulted his catering abilities, commenting on how little beer was available,

2 Women who had never married were represented by their fathers, or by a stand in for the father.

and accusing him of having hidden some of the liquor from them. They gave liberally of the beer that they had allocated to themselves, inviting others to come and sip from the beakers, but took some time before formally allocating beer to the visitors. Both hosts and guests deposited the residue of their heavy smoking and drinking in the vicinity, littering the hut with spent matches and pipe ash, spitting copiously on the floor (their home-made pipes had no filters) and frequently going out to urinate next to the garden or byre. When all the liquor was finished they would leave, having made sure (or so they thought, or pretended to think) that any hidden liquor had been produced and consumed. An interesting thing about all this was that both hosts and visitors would claim that the homestead and its inhabitants had now been *blessed*, their welfare secured, at least until the next time beer was brewed.

This is, of course, probably a fairly ordinary example of the kind of thing that tends to confront anthropologists in the field, and which they have to try and make sense of. In my case, I had come to the Transkei to study migrant labour and the effect it might have on traditional religious practice, specifically on ritual. I supposed that the absence of many men for a large part of the year as migrant workers would impact (negatively) on the way in which the ancestor religion was practiced. It was in pursuing my research along this line that I found myself having to attend beer drinks to make contact with people. Not that this helped my particular line of inquiry, because I was soon told that asking questions about the ancestors at a beer drink was not appropriate. However, what did emerge from regular beer-drinking encounters was both disconcerting and welcome, because I discovered that there were a number of beer drinks directly related to the process of going out to work as a migrant, and of returning home again. These turned out to be fully religious events, rituals aimed at securing the blessings of the ancestors for the forthcoming journey and the time at work, and they were performed along with a number of other ritual actions that had these goals. And on return from work, another series of ritual actions took place, including, sometimes, a large beer drink called *umsindleko*, the main purpose of which was to give thanks for the migrant's return, interpret his time at work in a socially appropriate manner, and re-incorporate him into the community. Far from migrancy making it more difficult to celebrate ancestral rituals, it provided instead an important focus and opportunity for the construction and celebration of ritual events. There was, in other words, a significant link between migrant labour and beer drinking that compelled me to change the focus of my research.

As time went on I found other significant links between labour and beer, which are documented in what follows below. Fieldwork in Shixini during the cultivating season was particularly trying. Sunrise saw many people already busy in their distant fields or nearby gardens, and the work of ploughing and planting and, later, weeding, consumed much of their time during the months of November through to March. It was difficult to find anyone to talk to during the day. This pattern was broken over weekends, when many ritual and beer drinking events, and the meetings of the youth associations took place.[3] Much of the work of ploughing and planting, and a good deal of the weeding, was done by groups of people who gathered together at the request of an individual homestead. Often, the receiving homestead was then obliged to brew beer to reward the workers for their labour. People who had not participated in the work party but who had been working elsewhere, perhaps on their own or with one or two associates, could also attend these beer drinks, where they had rights to beer. So in the mornings from sunrise at about 5 a.m. until mid-day, people were busy in the lands, in the afternoons they attended the beer drinks associated with agricultural work, and at night they went to sleep with the sunset. For weeks on end there seemed to be one work party, and one beer drink, after the other. I came to the (by this time not so reluctant) conclusion that successful fieldwork would mean acquiring a taste for the yeasty, thick and nourishing maize beer, and an ability to spend hours sitting on the floor in smoke-filled huts, recording what went on at beer-drinking events.

---

[3] The Xhosa youth associations have been described by Philip and Iona Mayer (1970). They continued to operate in Shixini until about 1980, when they ceased to exist due to a variety of factors (see McAllister and Deliwe, 1996).

What emerged from my observations during the cultivating season was that there were various ways of getting agricultural work done, and a wide variety of different types of beer drinks, some of which were directly linked to agriculture and the organisation of agricultural work. Longer residence also revealed that there were many other beer drinks, not obviously related to labour of any kind. Some of them were obviously religious and seemed to be substitute rituals for what would normally have been an animal offering to the ancestors. Others were held to mark events such as the passing of a swarm of bees that had temporarily settled in a hut – the bees being perceived to be manifestations of the ancestors, and referred to as 'honourable guests'. If you have distinguished visitors, you ought to feed them, and once the 'guests' had left, the community was called in to do justice to the brewers' efforts.[4] Another beer drink was held to give thanks to the ancestors for the harvest, though as far as I could see at first it had no specifically religious content – not even words addressed to the ancestors. This perception was to change later.

Building the homestead

Like 'beer for harvest' most Xhosa beer drinks were not overtly religious at all, yet they were of such obvious significance in the daily lives of most adults in Shixini that I spent a considerable amount of time attending them, recording what was said and done, and writing about them. In my first attempts to understand the significance of beer-drinks I was informed that they were a way of 'building the homestead' (*ukukwak' umzi*). Over the next few months I heard this term in a variety of other contexts also – a migrant worker who returned home after the perilous trip to the mines on the Witwatersrand was said to have been out to 'build the homestead'. Killing an ox or a goat for the ancestors in order to propitiate them, or to secure the ancestral status of a recently deceased grandfather, too, was an act that 'built' the homestead. Eventually it became clear that the notion

---

4 This relates to what are called *izilo*, which are manifestations of remote clan ancestors in animal form. Every clan is associated with a number of such animal spirits, which are particularly important in divination and shamanism. Bees that swarm into a hut are regarded as manifestations of the *izilo*.

of 'building' entailed a combination of social, economic and moral-religious elements, and that understanding life in Shixini would entail an analysis of how these elements were interrelated.

I concentrated at first on the beer drinks which were linked to migrant labour, particularly the large *umsindleko* beer drink sometimes held to welcome back a migrant after a spell at work. A short monograph on this beer drink was published in 1980, and I spent the next few years analysing material on other beer drinks and their associated oratory, and in writing this up in the form of a PhD thesis, completed in 1987. It has taken some years for me to get back to these initial fieldwork impressions and to the relationship between beer drinking and agriculture, including co-operative work. The significance of beer drinking to the productive efforts of each individual homestead, to the establishment of community and to the overall ethos of mutual assistance was clear from early on, but the details have not been spelt out before.

In the chapters below I proceed along the following lines. First, a brief introduction to Shixini is provided, by giving basic administrative, demographic, ecological and economic information on the fieldwork area. This section also includes an introduction to the people, in terms of some of their social and ideological characteristics. Since this book is largely about rural labour and agriculture, it is necessary to place the reader in a position to assess the ethnographic material in terms of the position of Shixini in the wider South African society, including its political and administrative history. I argue that the maintenance of homestead production in Shixini can be properly understood only if this wider context, both its contemporary nature and its historical development, is taken into consideration. In other words, although Shixini may be a relatively isolated rural area, it has for long formed part of a wider context in which the dominant feature has been participation in a modern state with a developing urban-industrial sector, but which was politically constructed in both the colonial and post-colonial periods in such a term. Agriculture and the labour processes through which it is sustained have to be understood within this context.

Photo 1
Shixini men and women at an *umsindleko* beer drink, held for a newly returned migrant worker

The ecological setting is also mentioned in this introductory chapter. Climate and topography are important constraints in any pre-colonial and pre-industrial subsistence system, and the nature of the pre-colonial Xhosa adaptation to the land is sketched. This serves as background for the significant changes that took place in the mid- to late-1800s, and which continued in the early 1900s as Xhosa speakers and what was left of their land were drawn into a wider political economy dominated by a system of racial exclusion and oppression. It is also important background for an understanding of contemporary agriculture in Shixini, especially cultivation, and for an understanding of the role of co-operative work.

The basic economic data indicate that Shixini people today, like many other rural people all over South Africa, depend very heavily on a cash income from the developed urban-industrial sector, or from the state in the form of old age and other pensions. But it is also evident that rural residence provides Shixini people with access to valuable agrarian resources and at least a partial subsistence. In particular, arable land is not in short supply, or at least the shortage is not as acute as in other places. It is argued that this has been possible partly because of Shixini's situation within a wider system, which in the last two decades has provided an alternative to rural residence for many of its people. Demographic data indicate that the rate of urbanisation is at least the same as the natural population increase, if not higher. Within Shixini, however, the great majority of homesteads still actively practise agriculture. It is important to many people not only as part of their existence as rural Xhosa, in other words as part of their identity, but also economically, contrary to what virtually all commentators on Transkei agriculture believe.

A consideration of agriculture in Shixini thus involves two related issues; an examination of the role agriculture plays in meeting local food requirements, and its importance in maintaining a particular identity. The second issue, concerning identity, is probably the easier of the two. It is well known that attachment to the land and involvement in rural activities, including agriculture, is related to a rural Xhosa identity of a conservative kind. However, the challenge is to establish whether this identity has any material substance. Is it backed up in any meaningful sense by subsistence production, or are agricultural activities in rural areas of mere symbolic value, demonstrating an attachment to the land that is of little relevance to daily subsistence? This is a more difficult matter, since it involves trying to establish figures for yields. We simply do not know how much maize (the staple crop) is produced in the former homelands (Beinart 1992; Bernstein 1996a, 132); all we have are estimates and the results of surveys that are of dubious validity. For other crops we do not even have this. In addressing the question of maize yields I take an anthropological approach, asking such seemingly ridiculous questions as 'What is a harvest?' and 'What is maize?' with unexpected results. My findings indicate that agricultural production in places like Shixini is higher than most people believe. Finally, I address the question of what the links are between these two issues, between rural identity and agricultural production, by asking how the *practice* rather than the ideology of subsistence production is related to questions of identity.

It is in providing answers to these sorts of questions that one has to venture into the realm of beer drinks and co-operative work. Given the unusual, even controversial, findings on agricultural yields in Shixini, the question of what makes it possible to practise agriculture in an area such as this becomes a crucial one, given the shortage of resources such as labour and implements, and given generations of neglect of rural production by successive governments in South Africa. This, in turn, is highly relevant to the question of rural development. The answer is not complicated; it is through co-operative labour that agricultural production is possible. But what sustains co-operative labour? How is it organised? What prevents its breakdown? What social institutions and ideologies support it? The answers to these sorts of questions *are* complex, and require a consideration of the nature of the homestead, the associated concept of 'building' the homestead, the general nature of the relationships between homesteads, and the general nature of labour in Shixini, given its situation in a wider South African context. This is considered against the background of the literature on co-operative labour in Africa in general, which is reviewed, and which serves as a point of reference and comparison, in this way facilitating the analysis of work groups in Shixini. The analysis, in turn, points to ways in which co-operative work in Africa in general may be better understood. In particular, the conventional typology of forms of co-operative work in Africa is found to have outlived its usefulness.

However, the key to unravelling the complex nature of co-operative work lies not so much in the literature on the topic. It lies, rather, in paying attention to one of the aspects frequently mentioned in this literature, but which does not receive any detailed analytical attention in it. I am referring to the fact that co-operative labour in Africa is often accompanied by and associated with festive or ceremonial eating and drinking of some sort, usually including indigenous alcoholic beverages such as sorghum or maize beer, palm wine, and the like. The key to understanding agricultural work in Shixini, it is argued here, lies in an understanding of the beer drinks that are associated with it, and of beer drinking in general. Beer drinking takes place in individual homesteads but, as indicated above, it involves groups rather than individuals, and is therefore of significance beyond any individual homestead. How homestead, agriculture, labour and beer drinking are connected lie at the heart of the analysis presented in this book, and form the subject of chapters five and six.

In chapter five, the historical conditions under which Xhosa work-parties became the key to maintaining rural production are outlined, given the nature of the ecological adaptation and the pressures to modify this. The ways in which co-operative work is manifested in Shixini today are described and analysed as an outcome of these pressures, emanating from the wider political and economic context that Xhosa people found themselves in after colonisation and to which they had to adapt, changing the nature of agricultural production as well as aspects of the relationships between homesteads. Understanding contemporary manifestations of co-operative work involves paying attention to the different organisational forms, their particular utility, and the moral and social basis on which they are formed, given the social, economic and ecological constraints under which contemporary agriculture is practised.

The role of beer drinks in the organisation of co-operative work becomes clear here, but the analysis is taken further in the chapter six, where an attempt is made to show how beer drinking and labour in general are related and how they form, in effect, a single system, of which co-operative work and the beer drinks associated with it are but a specific manifestation of a wider set of principles. This involves an analysis of beer drinks in general, as well as an application of this analysis to the particular beer drinks associated with co-operative work. One of the main theoretical implications of this is that in an area such as rural Transkei, processes of production need to be understood in terms of the acts of consumption that accompanies them. At the same time consumption, a major pre-occupation in anthropology over the past decade or more, needs to be understood in the context of production. This is ironic in view of the fact that consumption studies were seen partly as a reaction against the concern with production that was prominent in anthropology in the 1970s. The concluding chapter summaries the main points of the analysis and tries to bring them together to illustrate how beer drinking and co-operative labour form an integrated means through which Shixini residents try to live up to the ideal of 'building the homestead', and in so doing attempt to act out and preserve a conservative rural identity.

Agrarian development in South Africa's communal areas

On the one hand, as indicated above, the analysis presented in the following chapters proceeds along anthropological lines. It seeks to present in-depth accounts of beer-drinking events and the organisation of agricultural labour. It searches for interconnections between different aspects of a segment of rural Xhosa society, and shows how agricultural production and other economic activities depend on aspects of Xhosa culture and social organisation as these have evolved in an attempt to meet the needs of current socio-economic and political realities. But this does not mean that the findings are only of some esoteric, anthropological interest. On the contrary, they are directly relevant to current debates about agrarian change, land reform and development in rural South Africa, particularly in the communal areas. This debate centres on recent policy announcements by the national Ministry for Agriculture and Land Affairs, which has been reconsidering land and agricultural policy under a recently appointed Minister, Thoko Didiza.

Some six years after the advent of an ANC led government committed, among other things, to the eradication of rural poverty, under the new Minister South Africa has recently entered what Cousins (2000, 1) calls a time of "fluidity and uncertainty as to the future of the ambitious programmes of land and agrarian reform" promulgated by the government soon after its installation and repeated in a variety of policy documents since then (e.g. NDALA 1998). Recent policy pronouncements seem to have relegated land rights and agrarian change to the sidelines, at least as far as the millions of poor rural black people are concerned, in favour of developing a relatively well off and small number of 'commercial' farmers. The 'subsistence' sector, it is feared, is on the verge of a new era of neglect, of being regarded as a sector where there is no potential for meaningful economic development, but merely capable, perhaps, of serving a 'food safely net' function. In reaction to the new policy direction, Cousins (2000) and others (e.g. Nkuzi 2000) argue that experience in other parts of Africa shows that this dichotomy between commercial and subsistence farmers is a false one. They argue that development initiatives and land reform aimed at bettering the lot of the masses of rural black residents, regarded by government and commentators alike as the 'poorest of the poor', is possible and desirable, and that it can make major contributions to the development of the economy as a whole. The new policy directions and the implications of this for the question of agrarian development and land reform, and the debate around this, are discussed in chapter three. I evaluate the new policy directions announced by the government in the light of findings from Shixini, and provide evidence to support the contention that it is misguided. I show that many of the people of Shixini, though poor, produce substantial amounts of food from their land. Also relevant here is the potential impact on rural development of the government's Wild Coast Spatial Development Initiative (WCSDI). A brief assessment of this initiative in terms of its possible impact on people such as those in Shixini, and its potential link with agricultural production in communal areas, is made.

My argument, however, goes somewhat beyond (or is also somewhat adjacent to) the current debate. Much has been written in the last few decades about the need for agrarian reform in South Africa, about the requirements for successful rural development and associated issues such as the need for infrastructural improvement and the provision of services, land tenure reform and its contribution to agrarian change, and so on. While getting the policy right is undoubtedly of vital importance, little attention has been paid to *how* rural people actually live their lives, about the nature of their communities as the contexts in which they exist, about how they organise themselves, how they relate to each other, and how this relationship is linked to the land and to the agrarian activities in which they engage. Rural development, land reform, infrastructural development, and other measures designed to produce agrarian change affect real people in real communities, living according to a roughly coherent patterns of behaviour, values, desires, and precepts. I try to show how we cannot understand the one without the other, how agrarian activities cannot be understood without looking at the social, organisational and moral context in which they take place and which play an enabling role in agrarian life. I am concerned with how people use their 'social capital', if you like, to produce economic value, and how they strive to maintain the social, moral and organisational basis on which their productive activities depend.

The implication of this is that policy and development activities aimed at addressing the economic position of 'subsistence' farmers, whether initiated by government or by other agencies, would benefit from a vision that recognises the value being produced from the land, and from flexibility, making provision for local realities to be taken into account, recognising local organisation, knowledge and skills, and enabling these to inform the provision of services and support for agrarian development. In the case of Shixini, this includes an understanding of how residents are able to maintain a sense of community and identity, and of how they are able to harness the socio-cultural resources at their disposal to engage in productive activity, with some success. This issue dominates the second half of the book, and its implications are re-visited in the conclusion.

2

Shixini and its people

Historical and administrative background

The people who inhabit Shixini and adjoining areas are members of what was, historically, and technically still is, part of the Gcaleka chiefdom, the senior branch of the Xhosa proper (Hammond-Tooke 1975). Like other Xhosa-speaking chiefdoms, the Gcaleka became subject to British colonial control when the Cape Colony annexed the area known as Gcalekaland, the present districts of Gatyana and Centane, in 1885. Other territories east of the Kei River were annexed between 1879 and 1887, and together with East and West Pondoland, annexed in 1894, were administered through a council system and fell under a single chief magistrate for the Transkeian Territories (Jackson 1975).

One administrative expression of this legacy lies in the creation of the Gcaleka Regional Authority, part of the Transkei's administrative structure from 1955, when 'traditional authorities' were established in terms of the Bantu Authorities Act. The Gcaleka Paramount, Chief Xolilizwe Sigcau, held the chairmanship of this Authority, which included the district of Gatyana, where Xolilizwe resides, and also the districts of

Centane, Idutywa and Xhora (Elliotdale), parts of which are occupied by people of Gcaleka origin. Shixini is one of the 39 Administrative Areas or wards, that were grouped into seven Tribal Authorities, and that make up Gatyana district, together with the town itself and its commonage. Shixini and an adjoining ward, Ntlahlane, make up Jingqi Tribal Authority, and the chairman of this Tribal Authority, Chief Mandlenkosi Dumalisile, is also Shixini's headman.[1]

Xhosa chiefs were almost invariably polygynous, as were many commoners, and their different wives and their children were placed in 'houses', which were ranked in order of seniority. First in the ranking was the 'great' house, and the senior male child of the senior wife in this house, the 'great' wife, in theory succeeded his father as chief. Second in importance was the 'right hand' house, with the senior son in this house usually being given a portion of the chiefdom to rule over on behalf of the senior chief.[2] Chief Mandlenkosi, Shixini's headman, is the genealogically senior living representative of the right hand house of Hintsa, paramount chief or king (*ukumkani*) of the Xhosa and head of the Gcaleka chiefdom in the early 19th century.

Shixini people refer to themselves as 'Gcaleka' from time to time (mainly in ritual contexts), and are conscious and proud of their heritage and identity. It should be pointed out, however, that the term Gcaleka gives a somewhat false impression of unity and exclusiveness, political and cultural, which does not actually exist on the ground. Gatyana and the other districts historically part of Gcalekaland contain many thousands of Xhosa people who are not historically Gcaleka but who are nevertheless culturally and linguistically difficult to distinguish from those who are, except, in some cases, in terms of clan name and historical origin (e.g. in the case of 'Mfengu' people who were settled in the area by British colonial authorities — see below). In addition, Xhosa speakers in Gatyana are culturally extremely similar to the other major historical Xhosa chiefdom, the Rharhabe, whose descendants live mainly west of the Kei river (in-

1 Ntlahlane has its own headman.

2 For a fuller discussion of the nature of chiefly succession, the different houses, and related matters, see Peires (1981).

cluding the former Ciskei 'homeland', now, like the Transkei, part of the new Eastern Cape Province), and they are also culturally and linguistically very similar to other nearby groups which were independent chiefdoms in pre-colonial times, such as the Thembu, Bomvana, Mpondo and others, the chiefdoms which make up what have been referred to as 'the southernmost Nguni tribes' (see van Warmelo 1974, 60; Jackson 1975).

From 1963 until the demise of apartheid in the early 1990s, Transkei was nominally self-governing and, from 1976, 'independent' in terms of the state's strategy of 'separate development'. With the advent of a new democratic government in South Africa and the dissolution of the homelands and their administrations, the Transkei became part of the Eastern Cape Province. Within the province a new administrative order has been developed, including District and Regional Councils, though how this will impact on the older structures is not yet clear. More recently policy has shifted towards giving urban municipalities a larger role in the districts in which they are situated, but again, how this will dovetail with the role played by traditional leaders is not yet known and is still (November 2000) being debated. In many parts of the Eastern Cape (primarily in the former Ciskei) chiefs and headmen are no longer as relevant as they were in the past, since they were sidelined in the 1980s by ANC-affiliated residents' associations belonging to SANCO (The South African National Civics' Organization). Residents' associations tended to see chiefs and headmen as conservative and representative of the old political dispensation.

Despite widespread support for the ANC in rural Transkei, civic associations did not made much headway in places like Shixini, and the chiefs and headmen of Administrative Areas are still paid government officials. The judicial and other functions of Tribal Authorities were suspended after the ANC government came to power in 1994, and Transitional Rural Councils (TRCs) were set up in districts like Gatyana, ostensibly to take on the powers that were held by Tribal Authorities. However, the Tribal Authority system remains in place and the judicial functions of chiefs, which allowed them to hear certain civil and criminal cases, have recently been restored. Land administration continues under the leadership of chiefs and headmen, and the relationship between Tribal Authorities and the TRCs still needs to be clarified and formalised. Like other aspects of local government in the Eastern Cape, it is currently in a state of flux.

Recently, the ANC led government has made formal provision for the political representation of traditional leaders, who retain wide appeal in many rural areas of South Africa, including many parts of Transkei, with the institution of a National House of Traditional Leaders, and in the Eastern Cape the regional government has established an Eastern Cape House of Traditional Leaders. At the present time, the Gcaleka paramount and other traditional leaders from the older Gcaleka Regional Authority are represented on the Eastern Cape House of Traditional Leaders, but the role that this House will play, and of traditional leaders in general, is still being negotiated.

There are ten sub-wards in Shixini, each under a sub-headman (*isibonda*) who sits on the Tribal Authority and acts as intermediary between the chief and the people. However, most Shixini residents do not attend the regular meetings of the Tribal Authority, and the chief seldom travels to the distant parts of the ward, partly because of its size and the distance involved. The sub-headmen therefore play a crucial role in maintaining communication between chief and people. Land rights are granted by the sub-headmen in collaboration with the senior men of the sub-ward, and formally ratified by the chief. An indication of the chief's role as manager of the land, and of his power in this regard, was given in 1983 when the long-established sub-headman of Folokhwe sub-ward, an active farmer, decided to take over a large, long-abandoned garden belonging to a homestead which had died out. He fenced the garden and commenced turning over the soil in preparation for cultivation, but before long the acting chief at the time (Mandlenkosi's younger brother) got to hear of this unauthorised use of communal land and paid a visit to the area to see for himself what was going on. The result was that the sub-headman was removed from his position and ordered to take down the fencing and cease his cultivation of the garden. Most of the time, however, the chief is not directly involved in sub-ward affairs, though he remains the ultimate arbiter of local disputes and manager of the land under his authority.

In Shixini, older people are relatively conservative and readily identify themselves as *amaqaba*. Literally this means 'those who use red ochre' (on their clothes and bodies), but more broadly it refers to a traditionalist orientation, and among more westernized and educated Xhosa-speaking people it is regarded as a derogatory term, meaning 'uncouth', 'uneducated' or 'backward'. *Amaqaba* are also referred to as *abantu ababomvu* ('Red people'), and Red way of life and values has been described as an 'ideology of resistance' (Mayer 1980). In *Townsmen or Tribesmen?* Mayer (1960) contrasted Red conservatism with another folk category or Xhosa sub-culture, the 'School' Xhosa, who were educated and western-oriented, ideologically opposed to Red Xhosa, and who chose to resist white domination by engaging with it on its own terms. This ideal-type dichotomy between Red and School Xhosa no longer enjoys much currency (see Mayer 1980). Education, the independent churches, consumerism and other forms of westernisation have all made heavy inroads in just about every part of the Transkei, particularly among the youth, and *ubuqaba* (traditionalist values and way of life) has consequently been undergoing a process of change since at least the mid-nineteenth century. Nevertheless, most older Shixini people may still be classed as rural conservatives. Among younger people in Shixini conservatism is not so apparent (McAllister and Deliwe 1996) and many have opted for a life outside of the area at or near employment centres. However, there are substantial numbers of the younger generation who value a rural agrarian lifestyle and are committed to long-term involvement in it (see below).

As described by Mayer, Red resistance to white domination was expressed through an adherence to tradition, rejection of western values and behaviour patterns, and a determination to maintain a rural lifestyle associated with the value of cattle and cultivation. Among older Shixini residents, as well as among the younger ones who choose to maintain an agrarian lifestyle and base, there remain strong elements of this kind of Xhosa conservatism, which is closely linked to their agrarian activities, as following chapters demonstrate. Rural conservatism among young and old is also linked to the fact that levels of formal education are extremely low (those in wage employment as migrants are almost exclusively in unskilled jobs) and the traditional religion flourishes. Christianity is represented primarily in the form of African Independent Churches, though churchgoers are still a small minority of Shixini residents. Some of these features of Shixini life are dealt with in more detail below.

Physical characteristics and settlement

Shixini is an area of approximately 50 square kilometres, some 14 kilometres long and between 2 and 6 kilometres wide. It is located in the rugged coastal part of the district, on what tourists know as the Wild Coast, and is climatically and ecologically typical of the Transkei and Eastern Cape's coastal belt. It is situated in picturesque broken terrain with many steep hills and streams running down to the Shixini river in the north-east and the Jujurha river in the south-west, rising from sea-level to around 300m 15 kilometres inland. Rainfall is around 1000mm per annum, with 70% of this falling between October and March. In terms of Acocks' classification the vegetation is primarily coastal forest and thornveld, but it is only in the river valleys that much forest is found, and grassland (mostly sourveld) dominates the area. Soils, poorly developed due to leaching by rain, are mainly Glenrosa, Mispah and Swartland over shale and sandstone (Andrew 1992; Whisken 1995).

Shixini is 30 kilometres (at its mid-point) from Willowvale town, the district's small administrative, judicial and commercial centre. The town, in turn, is 32 kilometres by secondary dirt road from Idutywa, which is on the main highway running through Transkei, and from where connections to Butterworth (59 km.), Umtata (119 km.), East London (180 km.), and other urban centres may be made. Cape Town is 1259 km away, Johannesburg, 971 km. These distances, together with the extremely poor condition of the road between Shixini and Willowvale town, make Shixini, like the Transkei coastal belt in general, relatively remote from the main urban centres. Public transport by bus was sporadic during the period of Transkei's self rule, and now no longer exists at all, making unreliable and expensive private taxi services the only way of getting to and from town for most people. In summer the rains often make the road between Shixini and Willowvale town unusable to all but four-wheel drive vehicles.

Like most other parts of the Transkei, Shixini shows evidence of decades of neglect by previous administrations. Facilities of all kinds are either absent or inadequate. There are three junior schools (Sub A to Standard 4) and one junior-secondary school (Std 4 to Std 7) in the ward, but no high school, to serve a population of some 4800 people. Most adults have, at most, only three or four years of schooling; many have none at all. Like the isolated nature of Shixini, this contributes towards the conservatism of the area as a whole. Medical facilities are limited to two clinics (one established quite recently) staffed by nurses, with occasional visits by doctors from town. There are two or three private practitioners in Willowvale town in addition to the clinics there, and a Community Health Centre has recently been built on the outskirts of the town.[3] The nearest hospitals are at least two-hours by car from Shixini, in the adjoining districts of Centane, Elliotdale (Xorha) and Butterworth (Gcuwa). Indigenous healers include a number of herbalists and diviners, and many Shixini people have some knowledge of how to collect and use indigenous plant remedies and treatments.

Willowvale town boasts the district's only post office, police station, government offices, and petrol station, as well as various specialist shops and services — hotel and liquor store, furniture marts, hardware, wholesalers, undertaker, and so on. The 7 small general stores in Shixini itself stock only basic necessities such as maize and maize products (samp, mealie meal), flour, sugar, tinned foods, patent medicines, household utensils, paraffin oil, some building materials such as windows and doors, and clothing. There are no public telephones, electricity, water reticulation or public water storage system and no sewerage system. The great majority of homesteads in Shixini do not have even pit latrines, and people make use of the bushy areas away from dwelling sites for excretory purposes. Relatively low population density (see below) and the presence of

---

3 It was not yet staffed or operational in mid-1998.

wide-ranging and scavenging poultry, pigs and dogs, ensures that this does not become a serious health hazard, though standards of health are low and indicators such as infant mortality are high (Simon 1989). Water is collected primarily from springs and streams, and some care is taken to ensure that toilet activities take place downstream from water collection points. However, the latter are frequently fouled by livestock, which use the same water sources when in the vicinity of homesteads and not out in the grazing areas. Housing is primarily of the mud-brick and thatch variety, but in recent years square houses with corrugated iron roofing have sprung up, and a small proportion of wealthier people have large, western-style houses. Most of the latter are in the area occupied by the chief in the central part of Shixini.

In most of the ten sub-wards, homesteads are widely scattered in the areas demarcated for residential use. Other areas within each sub-ward are set aside for fields and for grazing. As indicated above, land allocation and control of other natural resources is based on consensus at the level of the sub-ward moot, consisting of all senior males, presided over by the sub-headman, and ratified by the chief. For new areas to be opened up as fields or as residential sites, for example, requires local consensus and the permission of the chief, which is given on the understanding that local consensus has been reached. Neither local people nor the chief have any authority over areas proclaimed as natural forests by the administration, nor over the seashore, which are the responsibility of the regional government.

In many parts of Transkei (as in the rest of South Africa) rural areas were subjected, in the period 1950-1980, to agricultural 'betterment' schemes. These have been well documented (de Wet 1989, McAllister 1989) and need not be discussed in detail here (their impact on rural agriculture and co-operative labour will be mentioned below). The basic idea was to relocate rural people into ordered residential areas, and to divide the land that remained into arable allotments and fenced grazing camps. A scheme of this kind for Shixini was approved by the Transkei government in the early 1980s but was only partly implemented before being aborted. However, this is another illustration of the power of the Tribal Authority over residents, since the betterment scheme was implemented with the TA's approval. The acting chief at the time had in fact initiated the move in the hope of generating 'development' for the area. At about the same time, the TA also gave its approval to the Transkei's Department of Agriculture and Forestry to establish a wattle and eucalyptus plantation in Shixini. The area chosen for the plantation was occupied by members of one sub-ward, who were made to relocate into one of the areas set aside as residential in terms of the betterment scheme, where they were allocated small (1,600 square metres), betterment-style sites. So the number of sub-wards in Shixini was reduced from eleven to ten. This new residential area was in an existing sub-ward, in which the chief lives and the tribal authority buildings, clinic and junior-secondary school are situated, and which was already more densely populated than the rest of the ward. Clearly, the position of the chief in relation to his people has important implications for the future of rural development, land tenure reform and the nature of local government (see Ch 3).

In most of the rest of Shixini, however, there was relatively little movement before the 'betterment' scheme was discontinued, and many of those who moved later returned to their old sites. Some were able to obtain certain advantages from residential relocation and the subsequent abandonment of the scheme. For example, people who did move as a result of the scheme established new gardens as part of their new homesteads, although these gardens were very small. However, many such people continued to use their old gardens as well as the new ones, at least while establishing the new one and making it productive. Once it became apparent that the authorities were abandoning the betterment scheme, the homesteads that had moved started to take steps to recover their lost assets. Some moved back to their old sites, others simply developed their new gardens to the desired size on their new sites and found ways of retaining their rights at their previous homestead (e.g. through establishing a son or a mother in it). In one or two cases, the move to a new site enabled the homestead to become considerably more productive than before, because it provided new opportunities for agricultural development through retaining both the new and the old sites and the gardens associated with these (see the case of Mntumkulu, below).

Ecology and demography

The defeat of the betterment scheme means that for most of Shixini people the scattered settlement pattern has been retained. It is a pattern that makes ecological and economic sense in that it takes advantage of the topography and the distribution of natural resources, given the absence of infrastructure and facilities such as water and sanitation supply, and it provides a modicum of continuity with the pre-colonial situation. It is important to consider the latter in view of the argument developed later that contemporary agriculture and co-operative work arrangements grew out of the changes imposed on Xhosa speakers towards the end of the colonial period, and which impacted on ecological adaptations at the time.

In the pre-colonial period, indigenous settlement in the Transkei was determined largely by the nature of the environment, which was characterised by 'small-scale repetitive configurations that contained a variety of natural resources' (Sansom 1974, 140). Each small area within a tribal territory was similar, containing within it the resources needed for subsistence according to the southern Bantu pattern — primarily arable land, grazing for livestock, perennial water, fuel, an area within which to hunt and gather, and access to resources such as thatching grass. Sansom has characterised this as the 'Eastern' (primarily Nguni) ecological adaptation, found right across southern Africa's eastern seaboard, between the escarpment of the Drakensberg and the sea, and he refers to the economy associated with it as one of 'contained investment' (*ibid.,* 135). Settlements were dispersed over the tribal territory (i.e. homesteads were widely scattered and not concentrated into villages) and subsistence activities were confined to the areas near the homesteads. The Eastern 'unit of exploitation' was 'a small, concentrated area, containing the full range of natural resources that individuals wished to exploit' (*ibid.,* 139). Economic relationships were similarly concentrated. The inhabitants of each local area pursued their subsistence activities more or less independently of those in other areas. As we shall see below, the contemporary organisation of agriculture and of co-operative work in Shixini follows this pattern to a certain extent, while at the same time being a response and adaptation to the position of the area within the wider socio-political and economic South African context and the historical factors impacting on it within this.

The 'Eastern' form of adaptation was associated with a decentralised political, judicial and administrative system, and with a high degree of local autonomy. Although each tribal area was under the overall authority of a chief, ecological conditions were suited to local regulation of resources. The chiefdom was divided into economically self-sufficient districts, under sub-chiefs or headmen, the boundaries between districts being natural markers such as rivers. But even within a headman's area, further decentralisation was possible, because 'repetitive configurations allowed definition of tiny areas associated with neighbourhood groups or kraals within a district' (*ibid.,* 140). In the past, such groups often consisted of close kin, under the leadership of a senior agnatic kinsman. Rights to grazing, arable land and other resources were allocated on the basis of membership of the local group, though ratified by the headman and, ultimately, by the chief. In many parts of the Transkei, still today, this local group is called *ummango* (plural: *imimango*), literally a 'ridge', and consists of a group of homesteads occupying a small local area. Homesteads were usually located on hilltops or ridges, often on land not particularly suited for cultivation, overlooking or near to their arable lands situated in the river valleys below, or on sheltered hill slopes (Kingon 1915; Shaw and van Warmelo 1981, 229-234). Typically, grazing was also found nearby, the cattle going out to graze each morning, and returning to the byres adjoining homesteads each evening.

Socially and economically, this produced a kind of 'inwardness' in terms of exploitation of basic resources (Sansom 1974, 141). Homesteads in the same local area were normally large enough to provide their own labour needs, but co-operated with each other economically when required, especially with regard to grain production. Between smaller homesteads, in particular, labour was offered on a reciprocal basis, though production was the business of an individual homestead. Livestock, too, were associated with a local group and herded collectively, though each homestead had its own particular animals. This helped to emphasize the unity and corporateness of the local group, which was often a kin group or based on a core of close kin. In some cases it may have consisted of a single extended family inhabiting a single, large polygynous homestead consisting of between ten and forty huts. There are clear resonances of this pattern of co-operation in places like Shixini today, as we shall see.

The crops grown included sorghum and a variety of pumpkins and gourds. However, milk was the staple diet rather than grain, and the environment was able to support a relatively dense livestock population. This made possible differences in wealth, and also a complex network of social ties and obligations based on debts in livestock, since cattle were needed for bride wealth and for other purposes such as ritual, transport and food. Conflicts caused by incidents such as cattle trespassing into arable land were resolved at the local level, between neighbours (though recourse to the local headman's, and ultimately the chief's, court was possible). Even the rituals associated with the seasons were celebrated locally and not synchronised at the tribal level, as with Sotho and Tswana speakers in the interior of the country.

In addition to the low-lying fields, some homesteads also had small, fenced gardens nearby, often on a fertile old kraal site, in which crops such as tobacco were grown. The fields, too, were fenced to keep out wild animals, but such fences were taken down at the end of the harvest, the lands effectively reverting to commonage, with cattle allowed to graze on the grain stubble in the winter months (Shaw and van Warmelo, op cit). Soils were relatively infertile, and old land was fallowed after a number of years and new fields established. This practice both maintained productivity at the desired level and prevented serious damage to the environment. From time to time homesteads split as sons moved out to establish their own homesteads nearby, sometimes on virgin territory. The political system as a whole was one 'geared towards expansion', and the sons of chiefs often broke away with their followers to establish a new chiefdom on new land (Peires 1981, 53). People were able to cultivate as much land as available labour made possible, the right to such land being obtained simply by breaking the soil (Shaw and van Warmelo, op cit.). Most of the agricultural work was done by women; men being involved primarily in clearing the bush off new lands.

To return to the contemporary situation, it is remarkable how much of the above description still appears to apply in Shixini, though we have to be careful not to mistake superficial appearances for continuities in life-style and well being, and to come to grips with the historical changes that have taken place and how they have affected economic life, in particular. These are looked at in more detail below in the section on the historical development of co-operative work. At this point it should be noted that in nine of Shixini's ten sub-wards people live in scattered homesteads rather than in village-type settlements. In certain parts of the ward, distinct clusters of homesteads in close proximity to each other are often found, but this does not change the basic principles of land allocation and land use. Gardens adjoin homesteads and may be as big as the homestead head desires, as long as the rights of neighbours and of other members of the sub-ward are not infringed. Fields are in separately demarcated areas, usually on the flood plains adjoining rivers and streams, from one to three kilometres from homesteads. Many fields, as we shall see below, have been abandoned altogether and replaced with large gardens. Sub-wards are divided into territorial sections or neighbourhood groups (*iilali, iziphaluka*) groupings consisting of one or more *imimango* (see above). These are informal groupings, which regulate their own affairs at the local level but are nevertheless recognised by sub-ward residents and the sub-headmen. These territorial groups or sections play a vitally important role in social and economic life, as we will see below.

Significant for the maintenance of agriculture is the fact that population numbers seem to have remained static in Shixini over the past two decades, at least. The number of homesteads and people in Folokhwe sub-ward, for example, has remained virtually the same over a period of 22 years. In 1976, when I first visited Folokhwe, it had 79 homesteads, and in 1998 there were 77. When Heron (1990) worked in Nompha in 1988 it had 67 homesteads, 9 less than in 1998. The increase is partly due to the fact that Nompha contained one of the new residential areas demarcated under the betterment scheme, and some of the people who established homesteads there (including three from Folokhwe) decided to stay after the scheme was abandoned. The population of Shixini as a whole, too, seems not to have grown much over the past twenty years. According to 1970 census figures published in Jackson (1975), the population of Shixini in that year was 3056. Transkei census figures for 1991 estimated Shixini's population at 4562 people, though this may have been based on extrapolations rather than a household census (Transkei 1994). According to the national population census of 1996, Shixini had a total population of 3965 people in 828 'traditional' homesteads and 50 houses, an average of 4.5 people, which seems low, given the current average household size in the three southeastern sub-wards of around six people per household. The latter figure, however, represents the de facto population. If absentees such as migrant workers are included at the rate of one person per homestead (Andrew 1992) the official census figure rises to 4793 in 1996. This is an average of 5.5 persons per homesteads (de jure), which seems more in accordance with other results for this and similar areas.

In a count of homesteads in the whole of Shixini on the basis of aerial photographs, Andrew (1992) found that there were 501 homesteads in 1942 and that this had increased to 585 in 1962 and to 836 in 1982, a rate of annual increase of 0.78% in the former period, and of 1.8% percent in the latter period. This higher rate of increase between 1962 and 1982 might be accounted for in terms of the difficulty of urbanising after a Nationalist party government passed strict influx control laws in the 1950s, and to the expulsion of many black people from urban areas during the 1960s and 1970s, forcing some Shixini people to return to their homes. However, the forced relocation of people permanently domiciled in 'black spots' and urban centres to the 'homelands', including Transkei, does not seem to have greatly affected population in Shixini, perhaps because it is relatively remote. It is possible, however, that there were more people per homestead in the earlier period (see discussion of declining homestead size, below), which could account for the rapid growth in the number of homesteads by the fact that younger people were establishing them earlier than before. It would also change the population growth rate. For example, if the number of people per homestead in 1942 was eight, this gives a total population of 4008; if the number per homestead in 1982, forty years later, was six, the total population then would have increased to 5016. Since then, it seems, there has been a decline in population if the figures are correct (i.e. 5016 in 1982 to 4793 in 1996). However, the number of homesteads has increased from 836 to 878, and it is unlikely that average homestead size has decreased significantly in this time. The low rate in the growth in the number of homesteads between 1982 and 1996, of only 42 homesteads, or 0.35% per annum, is probably accounted for by increasing urbanisation after the mid 1980s when it started to become easier for black people to leave rural areas permanently.

As mentioned earlier, much of the fieldwork for this study was done in three of Shixini's sub-wards, namely Nompha, Ndlelibanzi and Folokhwe (Map 2). In 1998 these sub-wards had a total of 1220 people, living in 76, 41 and 77 homesteads respectively, an average of 6,3 per homestead.[4] In these three sub-wards, 54% of homesteads are headed by men, 46% by women. Of the latter all (except for two divorcees) were widows, either of men who had lived in Shixini, or of men who had lived in other areas, the widow returning to her natal home after the husband's death. In addition to the substantial number of female household heads, women are in charge of the daily running of the homestead in cases where a male head is absent as a migrant worker (Heron and Cloete 1991).

A characteristic of these sub-wards, and probably of most of Shixini, is the relative stability of the population, not only in terms of size, as indicated above, but also in terms of mobility and the absence of immigrants. This stability is of significance in understanding the nature of labour in Shixini, which depends on high levels of trust and reciprocity. The people who live in Shixini have, mostly, lived there for a long time. Seventy percent of homesteads were established more than 20 years ago, and only 19% have been established in the past ten years. Current residents are the descendants of others who lived on the same site or nearby— over 80% of

4 Figures such as these are approximate because the number of homesteads changes constantly and it is not always clear whether a homestead without occupants has died out or is still viable.

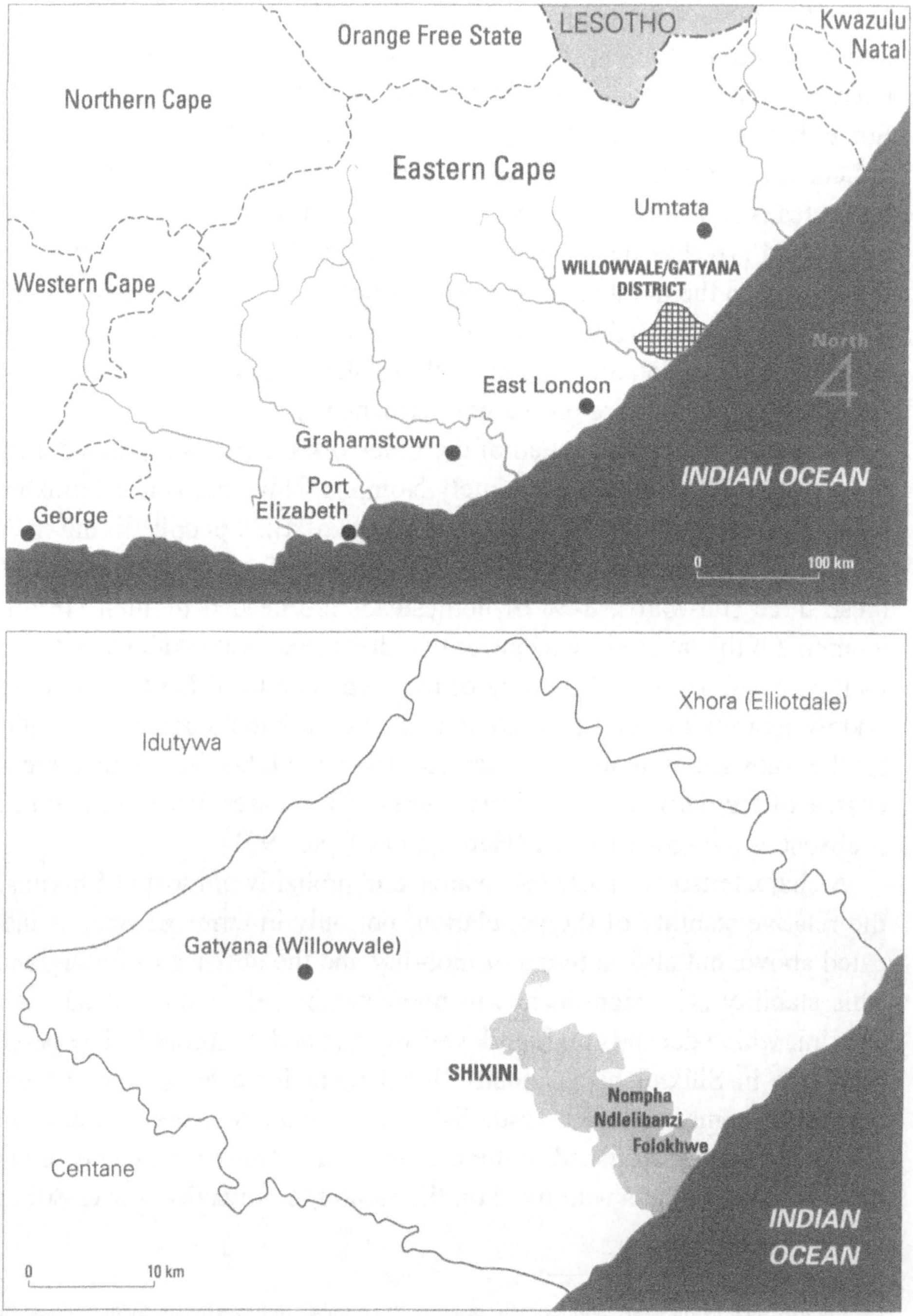

Map 2 Shinixi Administrative Area, Gatyana District

homestead heads' parents lived on the same site or in the same sub-ward as the current occupants and an additional 9% had lived close by elsewhere in Shixini. In only a single case had the parents of a current homestead head lived outside of the district of Gatyana. This stability shows little sign of change — all but one of the new homesteads built within the last 10 years were established in the same sub-ward as their parents' homesteads, and women who leave to marry still go primarily to homesteads in the same or nearby wards and seldom marry into another district. This level of continuity and stability is important in understanding the social and economic characteristics of the area. It is probably unusual in the context of rural South Africa, many parts of which underwent social disruption during the apartheid years, but whether it is unusual for coastal Transkei or not is uncertain.

What does seem evident, on the basis of a long association and familiarity with many of the households in the three sub-wards mentioned above, is that many of those who have chosen to leave Shixini for urban areas and other employment centres are the youth or young adults. Usually these are men or women in paid employment, who have chosen to live at or near their places of work on a permanent or semi-permanent basis, often together with spouses and young children. Some of them are likely to return to Shixini only if they find themselves unemployed, or in their middle age, if at all. Others, probably a minority, hope to build up some capital and then to return to an agricultural lifestyle in Shixini. A number of the case studies provided in the next chapter demonstrate aspects of this process, which seems to be an ongoing one, but which can take a variety of courses. In general, however, a certain proportion of the youth who are away at work as migrant labourers return to Shixini after a number of years, usually after they have married and had their first children, to engage in agrarian activities, either by choice or because of sudden unemployment and/or disillusionment with urban areas. Given the growing problem of unemployment in South Africa, it is possible that this trend will increase, though is likely that a significant number will prefer to remain in urban areas whether they are able to find work there or not.

Making a living

Although the emphasis in this book is on agriculture and the labour associated with agricultural and other productive activities, no Shixini homestead can survive without some form of cash income, and very few are able to secure access to cash by selling agricultural produce. Of the total of 194 homesteads in the three sub-wards surveyed in 1998, 80% had access to some cash income, usually in the form of migrant labour earnings and remittances, or social and disability pensions. The remaining 20% had no access to cash whatsoever, but some of these were able to acquire small amounts of money from time to time in various ways, as described below. For Shixini as a whole, the 1996 census results indicate that 27% of homesteads had no cash income.

Like other rural South Africans, Shixini people depend heavily on goods and services purchased with cash, but the importance of cash income relative to income in kind is not always easy to gauge. Pensioners say that they are able to meet many of their needs through the cash they receive but that they would have difficulty in managing without their garden produce. A would-be migrant worker lamented that he needed a job badly as he was presently surviving 'on maize alone'. Agricultural produce is important for daily subsistence, as we shall see below, but the land can provide, at best, all of a household's vegetable and grain requirements, possibly a small surplus, and, at worst, something of a fallback in hard times. Shixini people need cash for a range of manufactured food products (sugar, tea, coffee, flour, tinned foods, etc.), as well as many other manufactured goods such as clothes, footwear, cooking utensils and agricultural implements, and they need money to pay for transport, medical and school expenses, and so on.

Opportunities for local employment in any formal sense are extremely limited. A handful of people have paid jobs in the general stores, the schools and the Cobb Inn Hotel at the mouth of the Qorha River on the border with Centani district. There is also occasional employment at the plantation or with the road gang, which carries out repair and maintenance work on the district's dirt roads. Campers and fishermen from outside Shixini provide a few, very temporary jobs in the holiday season. Apart from local teachers, who are almost exclusively from outside the area, the only people in formal local employment in 1998 in the three sub-wards surveyed were one man who was employed as a night-watchman at the hotel, and another who was a policeman in Willowvale town. Both visited their homes over weekends. Self-employment is usually part-time, and included one owner of a small van, which he used as a taxi, and one man who owned a tractor and who hired out his services at ploughing time. A handful of others worked as thatchers or builders of huts or livestock-byres, mainly on a part-time basis, and there were three or four herbalists and diviners, none of whom made a regular living out of their calling.

There is a wide variety of ways of obtaining some cash within Shixini, but very few of obtaining a cash income which is sufficient for a household's subsistence needs. In other words, outside of migrant remittances and pensions, most cash income is irregular and small. One of the most common methods is the purchase and re-sale of alcoholic beverages, or the brewing and sale of a commercial alcoholic beverage locally called *mangumba*, but this is largely used only as a way of making a little money in an emergency. The costs are high (one has to travel to town to buy the liquor) and most people dislike the idea of their home being seen as a liquor outlet. Only two homesteads in the three sub-wards, one in Folokhwe and one in Nompha, sell liquor on a regular basis. Maize beer may also be made for sale, again on an irregular basis, since moral and social norms discourage it. The only socially acceptable way in which it may be sold is through the convention of a beer drink known as *imbarha*, at which a good portion of the beer must be given out free of charge. It is regarded as a means of obtaining cash in an emergency, and not as something that should be done on a regular basis. Nevertheless, there are two or three female-headed households that seem to brew *imbarha* more frequently than others, but this frequency is too low (about twice a year), and the amounts of money made too small (around R100-R150[5] each time) to be regarded as a regular means of obtaining cash income.

---

[5] Amounts in South African Rands refer to the period of fieldwork in 1998. One Rand is the equivalent of approximately US$0.14

A dozen homesteads sell manufactured commercial foodstuffs and other goods on a regular or irregular basis, and some of these are known as small informal shops (*iziroxo*). The range of commodities involved and the income derived is very limited in all cases. For example, Nobongile sold only potatoes and cabbages (purchased in Willowvale town) as well as paraffin. Only two of these households sold more than three or four commodities, and even here the range was limited to items such as paraffin, long-life milk, soap, sugar, beans and candles. Two women make occasional trips to Durban where they purchase second-hand clothing for resale back in Shixini. Neither makes anything approaching a full-time living out of this activity, however.

Other common ways of earning an occasional cash income include selling poultry or the meat from slaughtered pigs, and hoeing or performing other kinds of labour (e.g. collecting firewood) for wealthier homesteads. Wage labour, however, is rare and poorly paid (R5 per day), and most extra-household labour is associated with reciprocal mutual assistance, as will be illustrated below. A handful of women practise a craft of some kind, making and selling clothes, woven baskets and sleeping mats. The coastline provides some people with occasional cash as well as providing valuable protein from shellfish at spring tides. A few women collect and sell oysters to Cobb Inn, a laborious and poorly rewarded business. Since this can only take place at low spring tides and not all women have the ability to pick oysters, it is not a very common means of obtaining cash. Another source of income from the sea is a certain species of sea-weed which gets washed up in large quantities and which is collected by women for sale to a merchant from East London who makes occasional but irregular trips to the area for this purpose. Fishing with rod and line is the preserve of many boys and a handful of adult men. The skills, time and equipment required to fish successfully make it difficult to make money out of this occupation, though the hotel will purchase freshly caught fish (and also rock-lobsters) at around R5 per kilogram.

Hut building provides for a range of more substantial income-earning opportunities. For the common mud-brick and thatch huts, a recognised builder is hired to construct the foundation and the walls for payment of around R200. The wooden roof-frame is made and fitted for around R60, and a specialist thatcher completes the process for a further R300. Builders and thatchers are exclusively men. Young girls of the neighbourhood are usually hired to make the mud-bricks, and they are paid R12 for every 100 bricks made (about two days' work). The grass needed for thatching may be purchased from one or more women or acquired through organising a work-party of women to cut it. An elderly woman said that she sometimes made R60 a week by cutting and selling thatching grass (three bundles per day at R4 each). Then there is the fitting of the door and window, plastering the walls with mud, and the construction of the floor and the drainage ditches. Members of the homestead may do many of these tasks themselves, hiring specialists only for building and thatching.

Certain income-generating tasks are indirectly related to agriculture. Three or four younger men in the three sub-wards studied specialise in the task of constructing livestock byres and garden fences. Circular byres are made by setting a number of stout poles into the ground and then interweaving thinner, longer branches (*amahlahla*) horizontally between these. When Nophatile hired one of these men to construct a cattle-byre for her, in mid-1998, she supplied the poles (which she bought from the plantation in the centre of Shixini), as well as the *amahlahla*. The labour cost for the construction of the byre was thus kept down to R100. The person she hired told us that if he had been required to supply the poles he would have charged a total of R200, and another R100 for the *amahlahla*. Another man who sold poles for byres said that he sold them in R75 lots.

Few people have enough livestock to sell off animals on a regular basis, but this is a source of money in emergencies even if one has only a few head. Prices are high (e.g. R200 for a ewe; R2000 for a young cow or bull) but buyers outnumber sellers. The buyers are usually young migrant workers who invest in livestock as part of the attempt to set themselves up for a future on non-migrancy. In good years, people with surplus maize sell some of their harvest to others. Sometimes a person who has reaped well makes it known publicly that he has maize for sale, and the idea of 'a person who sells maize' is linked to this, as a public and known event. More often than not sales are relatively private and unpublicised. As with livestock sales, this is linked to the problem of witchcraft and the fear that envious people may, wittingly or unwittingly, harm those who are wealthy. Masilimane, who sometimes has surplus maize, told us that he occasionally sold a bag or two to anyone who came to ask to buy maize. His price was considerably lower than the price of maize bought at the shop, and he never sold less than a bag at a time. If smaller amounts were required he provided the maize as a gift, or sometimes (as we shall see below) in exchange for labour at a later date. It would appear that transactions involving maize are therefore not frequently ones that involve cash, with maize given and reciprocated for in the form of labour or in other ways.

Some also sell other garden produce such as cabbages, sweet potatoes, green maize on the cob, spinach, and the like. One Saturday morning in 1998 we observed Nophindile, a neighbour and pensioner who spent many of her younger years in the city of East London, set up a small table outside her homestead, on which were placed plastic bowls full of green maize and sweet potatoes for sale in the manner of a city hawker. An active and skilful farmer, Nophindile boasts a productive garden, from which she regularly sells quantities of vegetables, especially silver beet, carrots and sweet potatoes, on request. Sheep-owners are able to sell wool to a merchant who visits the area from time to time. A large bag of wool fetches around R100. Masilimane, who owns some 40 sheep, had three large bags of wool in his store-hut in mid-1998, awaiting the arrival of a buyer.

Migrant labour

As indicated above, the primary source of cash in Shixini, apart from pension payments, is in the remittances of migrant workers. Some 60% of homesteads had at least one migrant worker in employment in 1998, and an additional 20% had one or members who wished to be employed as migrants, some of whom were away seeking work. Many of these were people who had recently lost their jobs, with 37 % of homesteads reporting members who had become unemployed due to retrenchment, dismissal, ill health or old age. Some of the unemployed, however, were new entrants to the labour market. South Africa has a high rate of unemployment, current estimates being in the region of 35-40%, with few prospects of reducing the figure in the short or medium term. The findings on agricultural yields in Shixini, and the relevance of these findings for rural development, discussed in the next chapter, assume added significance in the light of this.

More than 80% of absentee migrants made regular visits home, some twice a year or more. Many migrants are said to work *eKapa* ('in Cape Town') but closer inquiry reveals that this term is used quite loosely to include small towns in the western and southern Cape, where many Shixini men are employed on fruit and dairy farms. Many also work in the fishing industry in places like Saldanha Bay. The majority of current labour migrants are men and boys and most of the male migrants who are married are not accompanied to work by their families. The pattern is thus largely the one established under apartheid rule, namely that of unaccompanied, oscillating, male labour migration. However, as indicated in the population figures mentioned above, many of the younger Shixini migrants have broken with this pattern in recent times and moved to town on a permanent or semi-permanent basis with their immediate families.

There is also a sizable and growing minority of female migrants. For example, Nokavala, who lives in Folokhwe, has four daughters living in East London. One works as a shop assistant, another runs an informal business from home, but the other two do not have regular jobs. The two who do earn send money to their mother every month, and write to her regularly, visiting home twice a year. Nokavala visits her daughters in East London about twice a year, especially if she feels unwell and needs to consult medical practitioners. She takes foodstuffs to them (e.g. maize) and in turn purchases things in East London that she brings home. In the past, she has made brooms for sale in East London, and she said that she intended to continue doing so. This kind of urban-rural interaction is fairly rare in Shixini, partly because distance makes travel to and from cities tiresome and expensive.

In the recent past, the terms that were used to refer to the workplace had negative connotations and were a reflection of the instrumental nature of migrancy. An absent migrant was said to be 'at war' (*emfazweni*), 'in service' (*enkonzweni*) or 'at the place of the white man' (*emlungwini*). The implication of such usages was that the workplace was strange, foreign and dangerous, and that entry into it was undertaken only to serve the rural home. One of the consequences of this attitude towards migrancy, based as it was on an appreciation of its ambiguous character, was an attempt to limit involvement in it to what was absolutely necessary, and a characteristic of labour migration in Shixini during 1976-1977 was that migrants of all ages spent almost as much time at home as they did at work (McAllister 1980). In more recent years (1980-1996) this has no longer been possible, as conditions of service on the mines and in other industries have changed (allowing men only short spells at home between contracts) and as jobs in general have become scarce. A partial exception to this appears to be the Western Cape fishing industry, where relatively long spells at home of around two to three months between labour contracts still seem possible. There is always a proportion of men at home, made up of the chronically unemployed, the temporarily jobless, the retired, the few non-migrants, and those on leave. And while many of those who have left Shixini for the urban centres with their families, and who are unlikely to return, no doubt do not subscribe to the old attitude towards migrant labour as a necessary evil, this attitude is still in evidence amongst men who retain Shixini as their home and residence.

Associated with this particular attitude towards migrant labour, as indicated earlier, is a series of ritual actions performed by and for migrant workers on their departure for work and again on their return home (McAllister 1980). The most elaborate of these rituals of labour migration is a large beer drink called *umsindleko*, at which a recently returned migrant is welcomed home and formally addressed by the elders and seniors of the community. In these speeches the migrant is praised for his success at work and reminded that the object of going out to work is to serve his rural home. He is praised for his efforts on behalf of his home and urged to repeat his performance in the future. The beer drink is also seen as an invocation to the ancestral spirits to continue to help and guide their descendents. Above all, the object of *umsindleko* is to ensure that the migrant interprets his migratory experience correctly — that is, as an end designed to serve the rural home and to reinforce the commitment to rural values. Such rituals help to make migrancy meaningful to people, by reinforcing the commitment to 'building the homestead'. In doing so it seems likely that the disruptive effects of migrancy are minimized.

3

Subsistence agriculture, or 'scratching about in the soil'?

Land and livestock

A visitor to Shixini is immediately struck by a general concern with cultivation and livestock. The landscape, together with observable everyday activities, suggests an agricultural lifestyle and gives the impression that Shixini residents practice mixed farming. The attachment of residents to the land is soon apparent, and the majority see themselves as rural farmers. These observations, however, thinly disguise the role of migrant labour and its importance in sustaining rural lives, the heavy reliance on cash through remittances or social pensions, the strong links with cities which some people have, and the permanent emigration of a substantial number of younger people to urban areas.

The cash contributions from migrant workers and from pensioners are crucial to sustaining livelihoods, but this does not change the self-image of most Shixini residents as rural farmers; on the contrary, it sustains it. Many researchers in rural southern Africa have illustrated the interdependence between cash income and agricultural production (e.g. Murray 1978; Spiegel 1980; Beinart 1992). The local view of migrant labour as a necessary evil justifies wage labour as a means to 'building' the rural homestead, where the term 'building' includes activities such as raising livestock and cultivating the soil. In addition, many migrants are still able to time their visits home to coincide with the period of ploughing, planting and the initial cultivation (November-January), and people who make visits to town time these to coincide with the period after the harvest, when there is little agricultural activity. The perception of migrant labour as the antithesis of an agrarian lifestyle is also blunted by the fact that many Shixini male migrants work as labourers in agricultural enterprises in the southern and western Cape. They are therefore at the bottom end of the migrant wage scale, with few prospects for improvement or occupational mobility, and it is not surprising that they look to their rural homes and communities for a sense of belonging and meaning. Migrancy notwithstanding, however, most of the homesteads in the three sub-wards of Shixini in which this study was conducted derived at least part of their livelihood, sometimes a substantial part, from agricultural activities.

All homesteads in Shixini have the right to gardens, which adjoin homesteads. Since most of the area mercifully escaped agricultural betterment (called 'planning and rehabilitation' during the era of Transkei's 'independence'), gardens can be quite large (up to 2 hectares, sometimes larger),[1] their size being limited primarily by the amount of labour and other inputs that the homestead has, as well as by the location of neighbours. In practice, although many homesteads also have fields, it is the gardens adjoining homesteads that are significant agriculturally. This is a relatively new development in the sense that as old, established fields (mainly in river valleys) declined in fertility and became unproductive, people adapted by developing large homestead gardens to take their place. The areas available for fields are very limited, whereas residential areas are still relatively large. Fields are in short supply, and no new fields have

---

[1] The average size of arable land holdings in the Transkei is 1.5 hectares (Porter and Phillips-Howard, 1997). Holdings in Shixini are larger than this if one includes fields, but since fields are hardly used this would be misleading.

been demarcated and allocated for at least twenty years. Nevertheless, the majority of homesteads can claim access to a field, but due to infertility through soil exhaustion most fields have to all intents and purposes been abandoned.

Only 8% of homesteads have no arable land, although every homestead is entitled to a garden. Eighty per cent of homesteads have a large garden adjoining the dwelling site, and a quarter of these have a small garden as well (see below). Twelve percent of homesteads have no large garden or field, but do have a small garden. Those without gardens, or with small gardens only, are often relatively new homesteads which have not yet accumulated enough resources (e.g. for fencing) to establish one, but which usually intend to do so in the future. There are also a few homesteads that have undergone decline following the death of the male head and the emigration of younger members, leaving perhaps only an elderly woman in residence, which no longer cultivate and have abandoned their gardens. And in a few other cases women who have returned to Shixini, leaving their married homes in other locations, have not established gardens alongside their new homesteads. Some of them lack the labour and other resources to do so, and subsist primarily on old-age pensions, others do not rule out the possibility of creating gardens in the future. Gardens are used regularly, primarily for the staple, maize, but a variety of other crops are also grown.

Despite the entitlement of garden land and the role that gardens have played in replacing fields, there is little doubt that there is less land under cultivation in Shixini (and in the district as a whole) today than in previous decades. All over the district large expanses of long-abandoned, terraced fields can be seen. This is partly because of a tendency in the past to move the location of fields from time to time (see Hunter 1936, 72), and partly (in the inland part of the district) because of betterment schemes, where new fields were allocated and old cultivated areas were made to revert to grazing. But discussions with Shixini residents indicate that thirty or forty years ago there was considerably more cultivation than is the case today, and an analysis of aerial photographs (Andrew 1992) bears this out: The number of hectares being used as fields in Shixini in 1942 was 1327; and this declined sharply to 650 in 1982. Over the corresponding period the area being used for garden cultivation rose from 30 to 327 hectares, and the total number of homesteads increased from 501 to 836, the sharpest increase being between 1962 and 1982, as indicated earlier.

The decline in field-use was at least partly compensated for by the higher fertility of gardens and the greater intensity of garden cultivation (McAllister 1989). The limitations on field cultivation include low soil fertility, distance from homesteads, and the fact that they were/are unfenced, making fields difficult to fertilise with manure from the byres, difficult to protect, and difficult to collect crops from on a day to day basis. Fields were used primarily for maize monoculture, at least in recent years as they became less fertile; gardens are much more suited to intercropping. As indicated above, some households have a smaller garden attached to the large one, indicating the status of the larger one as a replacement for the field. Some use the small garden exclusively for tobacco *(nicotiana rustica*), others use it for a wide variety of vegetable crops, yet others to provide maize during the period when the main crop (in the large garden) is not yet ready. This small garden is called *iqal'itye* (lit: 'first eating') or *mvelidinga,* which is hard to translate, but means something like 'ever present' or 'always trying'. Sometimes this smaller garden is not fenced with wire but with woven branches (*amahlahla*), like a byre. In such cases it may be distinguished from the main garden by the term *igadi yamahlahla* ('the *amahlahla* garden').

Like the main garden, the location and boundaries of the small garden are constantly moving to take advantage of micro-environmental changes. For example, livestock byres often adjoin it and if the byres are moved or re-built, small but highly fertile areas can become available for cultivation on their former sites. The small garden is also sometimes established on a former homestead or hut site, also areas known to be fertile (cf. Hunter 1936, 87). As old huts (as well as old chicken coops, pig-pens and livestock byres) collapse and new ones are built to take their place, so a homestead slowly changes its location, and the garden and *mvelidinga* are modified to take advantage of this. Indeed, often the motivation for slightly changing the location of the homestead is to allow for this expansion of the garden or *mvelidinga*.

Rights of residence are automatically also rights to use the grazing and other resources of the sub-ward in which one lives. Most homesteads own livestock of some kind,[2] but a significant number (35%) do not have any at all. The average number of cattle per homestead in 1998 was 3.5 head, or 6.3 for those homesteads having cattle. This is slightly higher than for the coastal half of the adjoining district of Xorha (ARDRI 1989) and for the Transkei as a whole, where the average cattle owner has six head (Porter and Phillips-Howard 1997). Sheep and goats are also not plentiful and, like cattle, they are unevenly distributed. The average number of goats and sheep per homestead is eight and six respectively.

While Shixini people feel that they require more cattle, the number that might be considered to be sufficient for a homestead is difficult to determine. The days when every homestead could have enough cattle to provide itself with a regular supply of milk are long gone. For one thing, homesteads are now much smaller than they used to be a hundred years ago (Beinart 1982) and current population levels in relation to available grazing would not permit such a large number of livestock. Similarly, for every homestead to have a team of six or even four oxen for draught purposes (in addition to other animals) would put unbearable pressure on grazing and other resources such as forest timber, which is used for byres, yokes and sleds. Currently, homesteads combine their cattle for purposes such as ploughing and transport and this seems to be an 'economic' solution in many respects, making the best use of available cattle without putting undue pressure on other resources. The various forms of combination and co-operation involving cattle are discussed in chapter five.

It is tempting, then, to see current cattle holdings in Shixini as about optimum in terms of the requirements of both agriculture and the natural

---

2 Livestock ownership is erratic and fluctuating. Disease and the use of livestock in rituals and bridewealth payments are the main reasons for the absence of cattle, sheep and goats from the byres of some homesteads; others may receive numbers of livestock in the future through, for example, incoming bridewealth, or the expenditure of accumulated earnings on a migrant's retirement.

environment, particularly pasturage, which is relatively good.[3] Cattle are the most commonly owned type of livestock; almost all ploughing is done by ox-draught and the only means of local transport is by ox-drawn sled. In addition, oxen are killed in all major ancestral rituals, goats in more minor ones, providing occasional large supplies of high quality protein. There is one tractor-owner in the area and his services are in considerable demand. However, it is costly to hire a tractor to plough, many of the gardens are on steep gradients, and the tractor in question is old and unreliable, frequently being inoperative for long periods.

In addition to cattle, goats and sheep, a few homesteads own horses, and most have pigs (usually between one and three) and small numbers of poultry. Both pig and poultry numbers vary considerably over time because they are frequently sold for cash or used to provide meat for domestic consumption, and they are important in this respect. Pigs, especially, are demanding in terms of labour, food and neighbourly relations (because of their tendency to ruin garden crops), and they are slaughtered regularly.

An agricultural ethos

One crop of maize, the staple food, is produced annually during summer. Ploughing and planting start shortly after the first summer rains, in about mid-November, although dry conditions may delay this until December and even, in exceptional cases, until January or early February. There is considerable variation across homesteads in this regard, and some may plough and plant as early as late September or early October. Few homesteads have the resources to plough and plant their gardens by themselves, and rely on their membership of co-operative agricultural 'companies' to get these tasks done. If one visits Shixini in December it soon becomes evident that people cultivate at different times, because some gardens have maize plants a metre or more high, while other gardens have not yet been planted at all. Even within a single garden, different areas may be

---

3 Botanical surveys in Shixini by students and staff of the Department of Botany at Rhodes University have demonstrated that grasses are, in general, not seriously overgrazed (Burchmore 1988).

planted at different times as a strategy to ensure a supply of green maize over a longer period than a single planting would give, or to take advantage of variations in weather conditions.

Once the maize plants have developed to a height of about 10 cm., roughly 3-4 weeks after planting, it is necessary to remove the weeds by hoe or, if planted in rows with a planter, by mechanical cultivator. It is vital to remove weeds at this stage if the maize plants are to develop properly. A second hoeing is required about eight to ten weeks later, depending on rainfall. Conscientious farmers may hoe a third time, at least in those patches where weeds and grass have developed most quickly. If there are two or three members of the household available to hoe, each period of hoeing may last for a number of weeks. Frequently, the homestead does a portion of the hoeing with its own labour and organises a work-party to complete the task.

Other crops are planted shortly after the maize germinates. This is labour saving because as these crops develop the necessity for weeding diminishes. Subsidiary crops such as pumpkins, melons, calabashes, cowpeas, beans, sweet potatoes, and others are planted in the spaces between the maize plants and in vacant parts of the garden, and also in the smaller garden. Maize and other vegetables are consumed directly from the fields as soon as they are ready, but the bulk of the maize crop is allowed to dry on the stalks before it is harvested in early winter (May-June). If there are no other crops left in the garden, livestock are then allowed to graze on the old maize stalks and to consume any sub-standard maize cobs left behind. The land then lies fallow until about August-September, when the soil is turned over (*ukugelesha*) in preparation for the next ploughing and planting. Many also fertilise the land with manure from their byres, which are conveniently situated next to the gardens. About half of Shixini farmers manage to cultivate a small winter crop, usually of peas, beans, sweet potatoes or cabbages (McMillan 1988).

The tempo and ethos of life in Shixini is primarily agricultural. In ploughing season attention is focused on the weather and crop prospects, on the amount of rain that has fallen or expected, and much discussion revolves around this and other matters agricultural. Preparatory activities include the repair of fences, the readying of equipment and making arrangements for co-operative ploughing. Social activities at this time also revolve around cultivation. No beer-drink or ritual may commence till mid-day, when those who labour in the land cease work, and many beer-drinks that take place are held to reward agricultural 'companies' for assistance with ploughing and planting. During the growing season (December to March) most people are pre-occupied with their gardens, with activities such as hoeing, planting subsidiary crops, thinning out maize plants, and setting their dogs on errant pigs. Livestock owners have to ensure that their animals are well tended and do not destroy crops, and many men and boys spend the days out in the grazing lands herding their animals, taking advantage of their proximity to forest and bush to cut poles or branches for building purposes, collect medicinal plants, and forage for edible wild-foods and other resources of use to their homesteads.

Harvest-time, too, brings an associated set of activities, as sleds are readied, granaries repaired and branches cut and collected for new granaries. The appearance and fertility of gardens evokes frequent comment and comparison. In May 1996, for example, people were excited about the large crops about to be harvested and many were busy constructing new granaries. Talk inevitably turned to those with very good harvests — the number of people required to help in the fields, the size of the granaries, the number of sleds of maize brought in from the garden, and the amount of beer that was going to be brewed later. Shortly after the harvest there is a flurry of beer-drinking activity as people brew 'beer for the harvest', to give thanks to the ancestors for what they have received, and to publicly recognise the assistance of those who did the ploughing and harvesting. During the fallowing period, gardens are extended, fences repaired, and manure transported from the cattle-byre to the garden and left there to be ploughed into the ground during the first turning of the soil in the new season (*gelesha)*. In the past this was also the time for building new huts, repairing or replacing livestock byres, and for rituals, including male circumcision. This is still partly the case but now many rituals are conducted around December-January, when migrants who work in town return for their annual holiday.

Although the overall ethos in Shixini is an agrarian one, and most people are involved in agricultural activities to some degree or another, as indicated above there are a small number of homesteads which have not used their gardens for many years, which do not have any livestock (and have not had any for some time), and which cannot claim to live from agrarian activities to any significant extent. They are usually headed by elderly women who depend largely on pensions or on the remittances of migrant workers, who visit their homes annually or less frequently. Of course, members of these homesteads often participate indirectly in agriculture, helping other people with hoeing, for example. And in the case of those with absentee migrants, they may engage in agriculture at some point in the future, given that many migrants intend to return home once they retire from work. In other cases they are homesteads apparently at the point of dying out altogether, although revival is often possible. In a few cases a homestead has no occupants, and stands empty most of the year, till the absent migrants return for their annual visit. In others, the head and wife are away at work and the homestead has a tenant who looks after it while the owners are absent. In some such cases the homestead members are minimally involved in agrarian activities, but the homestead still has cattle and goats, and its garden is used annually. The case studies given below illustrate some of the many and varied paths that may be followed in the course of homestead development and its involvement with agricultural and other subsistence activities, as well as the manner in which homesteads decline, revive, and disappear. The case studies are selected to provide a cross-section of Shixini people in terms of age, gender and migrancy status.[4]

1. **Nduna** was first interviewed at his homestead in Ndlelibanzi in 1977. He was about 50 years old, with a 22-year-old daughter, a 12-year-old son and three other younger daughters. He was still an active labour migrant, but nearing the end of his migratory career. On his previous return from work (in January 1977) he had purchased wire netting, a metal grain tank, six bags of maize and clothing for members of the family. He

---

[4] Pseudonyms have been used here and elsewhere to ensure confidentiality.

owned 14 cattle, about 15 sheep, and two horses. He was regarded as relatively wealthy. His garden was used annually, though his field had been abandoned. In 1976 he had harvested, he said, 4 bags of maize (though the real figure was probably twice this — see below). He was a member of a prominent and powerful lineage segment of the Ntshilibe clan, and belonged to a ploughing group headed by Ndabankulu, his FBS, and one of the most respected and wealthy men in the area.

By 1990, Nduna's son had started to go out to work as a migrant, and his household had expanded to include two granddaughters and a brother's daughter's daughter. His eldest daughter had married and was living with her husband in Mandluntsha, a nearby ward. He still had cattle and sheep, and now also goats, but the horses had died. The garden was still used regularly.

In 1992 Nduna died. In 1997 the homestead still had the garden, and it was used regularly, but there were no livestock at all. The son was still a migrant; he had married (accounting for the departure of most of the cattle as bride wealth to his father-in-law) and his wife and child were with him in Cape Town. He returned home regularly and sent money to his mother, No-amen, and No-amen now also received a pension. The daughter in Mandluntsha had returned home with her four children because her husband had died. Nduna's father's brother's daughter, was now also a resident. She was about 50 years old, and she had come to live at Nduna's homestead with her daughter of 15 years. They had left the homestead of her father, Phelephele, about 100 metres away, after the death of her elder brother, Sinotwana, and that homestead had now died out altogether.

It is possible that Nduna's son owned livestock but that these were kept elsewhere and not at his home, as is common with migrants. It is also highly likely that, with a number of girls in the home (some of them teenagers), that livestock will be coming in as bride wealth in the near future. So although the current position seems to be a heavy dependence on migrant remittances and pensions, the household contains a potential for expanded agricultural activity, particularly as it has a reasonable amount of (mainly female) labour. On the other hand, should Nduna's son decide to settle permanently in town with his family and his mother die, the homestead is likely to go into decline and possibly disappear altogether.

2. A younger member of the same Ntshilibe clan segment, **Mbuntuntu**, was a young man of about 23 years of age who had just left his Folokhwe home and gone out to work for the first time in 1977. His father, Mlulame, had died, and he had two younger brothers, one of whom had disappeared to town some years previously — no one knew his whereabouts, and he had neither written home nor sent any money. The other brother was still too young to go to work. His mother appeared fairly destitute. Her yields were 'nothing', she said, and she depended for maize on her natal home at Fumbatha, a nearby sub-ward, and on the generosity of her neighbour, Gunyuvu. Her last yield from her small garden, she said, was only half a sled-full (about 160 kg.), and the year before that it was even less. She had no oxen with which to plough, and the ploughing was done by Ndabankulu, her late husband's brother. The homestead had no livestock at all. At this point the homestead looked like one that might be on the verge of dying out unless Mbuntuntu and his brothers made some sort of commitment to a rural lifestyle, and it seems that they eventually did so, with the help of Ndabankulu.

By 1990 Mbuntuntu had married and was now regarded as the head of the homestead. He had four children (two months to seven years), and he returned home regularly from his job on a mine in Welkom. The homestead's garden had been extended and it was used regularly, still with Ndabankulu's help. They still had no livestock except for pigs and chickens.

By 1998 the homestead had expanded and was relatively well off. They had added a smaller vegetable garden to the large garden, built an additional hut, installed a new grain tank as well as a new homemade granary (*idladla*). The number of his children had increased to six and he now had cattle as well as goats (a large and well stocked byre was evident). He had lost his job in Welkom and had been unable to find a new one. One of his brothers, Luntu, had established himself in a homestead nearby which had belonged to Thyobeka, of the same lineage segment, and who had died, leaving his homestead vacant. But Luntu, too, had died suddenly and his wife had returned to her natal home. Mbuntuntu now maintained Luntu's homestead and had access to the garden, which he used annually. He had abandoned the homestead's field some years earlier, but now was able to use two gardens. In addition, Mbuntuntu and his wife run a liquor outlet, which brought in cash on a regular basis.

3. In 1977 **Masilimane**, a recently married young man, had no livestock and was working in Cape Town. His wife, MaDlamini, and their two small daughters were with him in the city due to illness, it was said, and they did not come home during the December-January period in 1976/77. It was said that they were 'School' people with at least some education. This turned out to be true to some extent, when I first met them a few years later. MaDlamini was one of the few Folokhwe women, at the time, who had taken to wearing 'School' type clothes, and when Masilimane accompanied me to a nearby *tshila* dance,[5] he turned up dressed in urban style (suit, hat and shoes). In addition, over the years it became apparent that they valued education very highly, and took pains to ensure that their two girls attended high school. Here, surely, were candidates for conversion from a rural to an urban lifestyle.

Masilimane's father's sister, Nowayilesi, a widow, occupied an adjoining homestead, where she lived with her two daughters. She had a garden, situated next to Masilimane's, but no livestock. Both Nowayilesi's and Masilimane's homesteads consisted of only a single hut. Masilimane had some livestock that were being kept by Kebeni, his older brother, but he owned no agricultural implements, and their garden did not yield much (they said that they reaped only 6 baskets of maize on the cob in the 1976 harvest, around 100 kg of grain).

Masilimane continued to migrate to work annually till 1983, but he ceased to do so in that year. He and his wife made some money for four or five years after that through liquor sales, but this was a temporary thing, and not on a large scale; they often ran out of supplies. MaDlamini sometimes collected oysters for sale to the local hotel, and Masilimane is one

---

5 *Tshila* was a public dance performed by male initiates while in seclusion after being circumcised. It was only rarely seen in the late 1970s and early 1980s, and has not, to my knowledge, been performed in the area since then.

of the skilled anglers in the area, so he was sometimes to able to sell fish to the hotel, and frequently able to provide it for the supper pot.

By 1987 their homestead had developed somewhat. They now had two huts, 15 cattle, and some small stock. The garden was being used regularly, and sometimes they also cultivated a borrowed field. The maize yield in 1986 was said to be 5 bags (about 500 kg.), mainly from his garden. There was 'virtually nothing' from the field because of infestation by beetle. By 1990 he had added goats to his other livestock, and now had an *isibhaya* (small stock byre) as well as *ubuhlanthi* (cattle byre). Innovatively, the garden was joined up with his father's sister's garden, and used as a single piece of land. In 1987 the two gardens were surrounded by a single fence but kept separate by a strip of unploughed land between them. By 1990 the strip had gone and the two were being cultivated as a single garden. This made sense given that Masilimane was the *usipatheleni* ('caretaker') for Nowayilesi and usually did the ploughing for her anyway. By 1990 he had acquired enough cattle to be a prominent member of a co-operative agricultural group or 'ploughing company' (see below chapter five), along with Ngcinibandla, a clansman and leading member of the sub-ward.

Today Masilimane is highly regarded as a successful farmer. He has effectively incorporated his father's sister's homestead and family into his — her single hut is now abandoned and she has her own new hut as part of his homestead, which now comprises a total of five huts; he has approximately 30 cattle, 40 sheep and 30 goats, as well as pigs and poultry. There are two small vegetable gardens in addition to the large one. He has a variety of agricultural implements, and his maize yields were more than sufficient for household requirements in 1996, 1997 and 1998. He no longer ploughs with Ngcinibandla, but is the leader of his own ploughing company, and he is *usipatheleni* (caretaker) at five other homesteads. His elder daughter is studying to become a teacher (her small child lives with Masilimane and MaDlamini) while the younger one has married and lives in the adjoining Xhora district. Four daughters of his father's sister are also domiciled in the homestead (in 1997-98 they were all away working or looking for work in Umtata or Cape Town), as are two small daughters of theirs. *Lobola* cattle that come in for these women when (and if) they marry will be Nowayilesi's but they will be managed with the homestead's herd. Nowayilesi now receives a pension, and this brings in necessary cash, but Masilimane is able to raise money also through livestock sales and the occasional sale of other agricultural products such as maize and wool.

4. **Nofinisi**, widow of Kebeni, Masilimane's older brother, provides a contrasting picture, but also one marked by development and change. Shortly after Masilimane 'retired' from migrant work in 1983, Kebeni went to Umtata to look for a job. Prior to this he had used his arable land regularly and also lived on his earnings as a part-time builder. He was successful in securing employment in Umtata, but it was very poorly paid. He seldom sent any money home to Nofinisi, who struggled to feed herself and her children. Their livestock died and the homestead deteriorated. It was only with help from her mother, in an adjoining homestead, and from Masilimane, official 'caretaker' of her homestead, that she managed to keep starvation at bay. After some years Nofinisi, together with their four children, decided to join Kebeni in Umtata. This was never very successful in ensuring a more secure existence for them, and she returned to Folokhwe from time to time, but without cultivating her garden. From about 1987 the homestead was inhabited only sporadically, and the garden was not used at all. To all intents and purposes it looked as if this was a family now fully dependent on urban wages, detached from agrarian life. In 1994, however, Kebeni died, and Nofinisi lost her foothold in Umtata and was forced to return home permanently. She and her children were desperately poor, but she eked out a living for a few years, sometimes returning to Umtata for short periods. In 1995 she hired a tractor to plough her garden for the first time in years, but it was unable to finish the job and she considered the R100 fee badly spent. In 1996 it was not cultivated. Finally, in 1997 Masilimane ploughed and planted the garden for her. He provided the seed, too. She was also able to establish a smaller garden below the main one, for cultivation of vegetables. When we visited her in February 1998 she was growing a sizable crop of maize in the garden, along with pumpkins, and in the smaller garden she had three types

of sweet potatoes, beans, spinach, cabbage, gourds and pumpkins, as well as maize.

Part of the key to Nofinisi's success in re-establishing herself lies in her kinship links with her mother, a mother's sister, and a father's brother's daughter, all of whom live nearby, as well as with her late husband's kin, particularly Masilimane. But another part of it is due to the fact that she had been actively engaged in hoeing for others in return for payment in cash or in kind (she also hoed for Masilimane in return for the seed he gave her and for ploughing her garden). This had enabled her to bring in some food and money during the growing period. Re-establishing the homestead also meant that she was able to become part of a neighbourhood network again. When we encountered her in her garden in mid-1998 she and her father's brother's daughter were replanting maize due to the fact that cattle belonging to Thekwane, a neighbour, had managed to break into her garden and cause some damage. We asked if she would ask for compensation from Thekwane and his wife. No, she said, she would not, 'because they are neighbours; they came to apologise, and they do help me from time to time'. She had obtained her sweet potatoes, for example, from another neighbour, MaMtshawe. Walking past this neighbour after speaking to Nofinisi we observed two women in the garden filling hessian bags with sweet-potato runners. They were from Chibi section, about two kilometres away. We asked them if they were purchasing these, and if they were related to the homestead. They replied in the negative on both counts, saying that they had merely heard that MaMtshawe had an abundance of runners, so they had come to ask for some and had been given. Each went off with a large bag full of runners, and we saw them repeating this the next day.

5. **Mntumkulu** was one of the people who were able to use the partially implemented betterment scheme to their advantage. In 1977 he was a young man of around 25 years of age living with his mother in his late father's homestead in the Ngingqi area of Folokhwe. It was probably the single most remote homestead in the sub-ward, situated in the extreme southwest corner. They were relatively wealthy, having inherited a sizable herd of cattle, sheep and goats from his father. A few years later he mar-

ried, and established a homestead next door to his late father's, where his mother still lived. For a short period he was actively engaged in cultivating both his own and his mother's garden, as well as the field which his mother and father had used for many years. By this time the threat of relocation was hanging over Folokhwe because of the betterment scheme. Mntumkulu probably saw this at least partly as an opportunity to acquire access to more arable land. He applied for a site for himself and an adjoining one for his mother in the new residential area near Ngadla shop and the local junior school. He moved to his new site, and left his mother at her old homestead where she could keep an eye on both of their original gardens. The field was no longer used due to poor yields from it. By 1996 the threat of relocation had disappeared but Mntumkulu had developed two large gardens at the new residential sites, one on either side of his homestead (technically one is his mother's, but she still lives in her old home). He therefore now has access to four gardens and, in addition, one of his sons has recently married and established his own homestead nearby, and Mntumkulu also farms the garden associated with it. He is an energetic man with access to great material wealth — in 1996-1998 his herds of goats and sheep numbered in the thousands, and his cattle in the region of 400-500. He regularly purchases vaccines and medicines to prevent and cure livestock diseases. He uses good quality maize seed and prepares his soil carefully. The results can be seen in the healthy, tightly packed maize plants, and his good yields. In winter 1998 he was growing cabbages in one of the gardens, and although they looked healthy he complained that they needed rain. He had been relying on barrels of water conveyed from the nearby stream by ox-sled, a time-consuming and laborious task but one potentially rewarding — there were approximately 1000 cabbages growing in the garden, and the local price for a medium-size cabbage was R2.50

6. It is by no means only middle-aged or older people who retain a desire to farm and a homestead in Shixini. The case of **Tatana** and **Nosindise** illustrate that there are also young people who wish to stay at home permanently, and who are using their earnings as migrant workers to establish the possibility of doing so in the future. In May 1996 Tatana,

Nosindise and their four young children lived in the Folokhwe homestead belonging to Nosisi. They had borrowed the homestead from her (Nosisi has another in Ndlelibanzi, acquired as a result of the betterment scheme). Tatana works in Umtata and comes home for holidays in December/January. He had been allocated a site nearby, down in a valley, and had turned the soil there. He intended to build his homestead there in the near future. I asked Nosindise if they did not prefer to live in Umtata. 'No', she said, 'We want to live here'. Why is that, I asked? '*Oh, sifun' ukulima, sifun' ukwakha*' ('Oh, [because] we want to cultivate, we want to build [the homestead]' she replied. This response is revealing in the sense that 'building the homestead' (*ukwakha umzi*), which Shixini people regarded as a moral and religious duty, is explicitly linked to 'farming' (*ukulima*). Eight months later we met Nosindise and a neighbour collecting sweet-potato runners in a garden some distance away from their homes (see above) and we asked her how their plans to farm were proceeding. They had managed to establish a small garden (*mvelidinga*) and had cultivated it for the first time. They had also borrowed a field from a kinsman in Ndlelibanzi, an adjoining sub-ward, and had ploughed it with the help of this kinsman's cattle when her husband was able to take some time off from his job in Umtata during November and December. This was the second field they had borrowed, she added, saying that the first one was not a success because the soil was infertile.

7. **Themba** is a young man of about 26 years of age. He has been working on a dairy farm near Plettenberg Bay for the last 7-8 years, and married in 1994. In 1997 he asked his mother to secure a site for him so that he could start to build his own homestead. She arranged a meeting with the Folokhwe sub-headman, ensuring that the discussion would proceed smoothly by presenting the sub-headman with a bottle of brandy at the time ('It was not a payment for the site', she said, 'it was just because there was a discussion taking place'). The site applied for was identified as that belonging to Themba's paternal great-grandfather and in this sense it was not a new site that he was requesting but permission to re-occupy and build on it.

The description of Shixini as dominated by an agricultural ethos should net be seen as a claim to some kind of unbroken continuity with the past, but merely as a description of the predominant type of economic activity. Similarly, the association between agricultural activity and local notions of *isiXhosa* (Xhosa custom or tradition) must not be interpreted as some kind of unchanging adherence to a particular body of 'traditional' thought and practice. As demonstrated in chapter five, below, the organisation of agricultural work in Shixini is itself a manifestation of important *dis*continuities with the past. Furthermore, although many older people in Shixini are strongly conservative, neither they nor their younger contemporaries are unaffected by changes and events in the wider environment. In the case of the youth, for example, the traditionally-oriented youth associations described by the Mayers have long gone, and have been replaced with new kinds of entertainment and activities such as football and *potsoyi,* an event associated with modern music, drinking and dancing (McAllister and Deliwe 1996). Most younger people are politically aware, and follow political developments such as current debates about the future role of chiefs and the Tribal Authorities. Yet the youth, too, are heavily involved in agricultural work. Dropout rates from the local schools are high, reflecting the involvement of children in a variety of household tasks, including agricultural work such as herding and tilling the lands (*ibid.*). Boys and young men, as well as girls and young women, play an important role in ploughing, planting, weeding and harvesting. Young men and women form the nucleus of certain types of work parties (see chapter five), and in this way make a vital contribution to sustaining agrarian activities. This involvement of the youth, together with the fact that there are enough younger people remaining permanently resident in Shixini to maintain the population level, is an important factor explaining the maintenance of agrarian activity, and one that needs to be taken into account in relation to questions concerning agrarian development, land reform and the future role and nature of local government, which we will turn to below.

Maize yields: What is a 'harvest'? What is 'maize'?

As the above case studies show, the aim of most Shixini homesteads, as far as their economic activities *in Shixini* are concerned, is to establish a degree of self-sufficiency through the practise of agriculture, including both animal husbandry and cultivation. As we have seen, livestock numbers are very low and take-off from this source makes a relatively small contribution to household subsistence. The case studies also indicate that agricultural yields are often very low, but this must be qualified by the fact that these were reported yields, not measured ones. Similarly, when homesteads were surveyed in 1998 and people were asked how much maize they had harvested from their gardens in the previous season (1997), respondents in southeastern Shixini claimed that they had harvested, on average, around 4 bags of maize (320 kg.) per homestead. Thirty per cent of homesteads, half of whom did not cultivate for lack of land or some other reason, said that they received 'nothing' or 'very little' from the land, 26% said that they harvested between one and four bags, a similar percentage that they had received between five and eight bags, and only 6% that they had harvested nine or more bags (i.e. 720 kg or more). This is very similar to the results of many other such surveys of maize production in the Transkei. If they are accurate one has to ask why people bother to cultivate at all, since these sorts of yields are hardly worth the effort and expense involved. The activities and aims of the homesteads cited above in the case studies would make little sense if gardens really produced so little food.

In this section I attempt to demonstrate, firstly, why it is in fact extremely difficult to obtain an accurate estimate of maize yields in a place like Shixini, and secondly, why figures based on reported yields, collected as part of household surveys or in other questionnaire-based ways, are utterly misleading. For one thing, it is generally known that Xhosa-speaking people tend to understate their yields when asked. Hunter (1936, 93) mentioned that Mpondo 'never boast of the crops they have reaped. They usually complain, when asked, of having a poor crop ...'. But there are also other reasons — some more technical, some cultural. They are worth considering in detail:

1. Arable land throughout the Transkei is irregular in shape and unevenly used, making it difficult to measure with any precision the amount of land under cultivation at any one time. Sometimes large margins are left between the cultivated area and the fence; sometimes only half the garden is used. In Shixini, a garden may have unused, rocky patches in it, or contain an old burial site that is never cultivated. This is not because people choose to bury their dead in the centre of the garden but, as indicated above, because of the dynamic relationship between garden and the rest of the homestead. A garden may be extended to include an old homestead site, including burial sites, but care is taken not to plough over the graves and in their immediate vicinity.

2. Inter-cropping: Maize is seldom grown alone and a variety of other crops occupy the spaces between maize plants. Inter-cropping makes measurements in terms of amount of maize per hectare difficult and meaningless. Looking at a field or garden when the maize plants have started to grow creates an impression that the land is badly farmed, especially when the broadcast planting method has been used rather than a mechanical planter, which produces neat rows of maize. There appears to be no order; some maize plants are crowded together, others widely spaced, and small parts of the garden may be devoid of maize plants altogether. The advantages of this are that it leaves plenty of space for secondary crops, such as cow peas, beans, sweet potatoes, pumpkins and melons, some of which may be invisible at the time of observation. Even *imifino* (wild greens), of which there are different varieties, may be found growing between maize plants, creating, to the untrained eye, an impression of sloppy weeding. This kind of cultivation makes a fair amount of sense, however. It provides variety, it is labour saving, it makes crop rotation unnecessary since a kind of random rotation occurs anyway, and it puts certain nutrients back into the soil (e.g. in the case of legumes). This means that soil fertility is prolonged and regular fallowing not needed, in turn saving on things like fencing costs and helping to conserve other resources such as grazing, since the total amount of land put under the plough is kept low.

3. Green maize cobs and other crops are gathered from the lands on a day to day basis for a considerable period of time before the harvest (this practise is called *ukufula*), so accurately measuring the amount produced requires daily observation. The period during which *ukufula* takes place may be as much as three months, depending on individual gardening strategies. Many people plant the section of the garden nearest the huts before they plant the rest of it, ensuring a supply of green maize over a longer period than would otherwise be the case. Sometimes three different plantings take place — two in the garden, one in the smaller garden — to guard against the risk of crop failure and also to ensure a longer supply of green maize. Even after the maize has dried out and ready for harvest, people continue to use it directly from the land, choosing the smaller and driest cobs (called *amaqhashu*), until the harvest itself.

4. The term 'harvest' is culturally specific, making it difficult to assess responses to a question like 'How much maize did you harvest in 1998?' There are a variety of reasons for this apart from that already mentioned: By 'harvest' Shixini people usually mean the amount of maize taken in from the fields at harvest time (around May-June), threshed and stored for later use. Sometimes people simply do not know how much they harvested, but usually they can give a figure of some kind. Some express the yield in terms of 'bags', but others only in terms of sleds or, if the harvest was very small, in terms of baskets. By 'bags' they usually mean an old-fashioned hessian bag filled to the brim, which they refer to as a 'Xhosa' bag — it weighs around 100 kg. in comparison with the new, shop-bought bags of maize, which weigh 80 kg. Sleds are home-made and vary considerably in size, some twice as big as others, so a figure in terms of sleds is meaningful only if one knows whose sled was used and what the equivalent in terms of Xhosa bags is (which those who regularly use the sled *and* who store their harvest in bags, know). The conversion is also one of maize still on the cob (in sleds) to grain (in bags).

Some homesteads (slightly more than a quarter) have metal grain-storage tanks, and a few still use pits dug in the cattle-byre floor. Most homesteads have a granary (*idladla;* pl. *amadladla*) made from long woven sticks (*amahlahla*), normally used for temporary storage only, the idea being to allow the maize cobs to dry out completely before removing the grain (*ukuxhontsa*) and storing it in bags or in the metal tank. During the two-three months of temporary storage continuous use is made of the maize. Tanks facilitate protection against damp and against weevil infestation (through the use of carbon disulphide, known and marketed as *iyeza letanki* — 'tank medicine'). If the harvest is small, however, it may never reach the tank, being consumed directly out of the woven granary for a few months until it is finished.

5. Local perceptions of what constitutes 'maize' (*umbona*) are more complex than one might assume. Harvesting is usually done by a group of women who strip the cobs from the stalks and the sheaths from the cobs, and empty their baskets-full of maize onto a number of points in the garden in readiness for the oxen with sled who transport it to the homestead. In the process the harvest is divided into two — maize (*umbona*) or 'good' maize (*umbona mhle*), and 'spoilt maize' (*umdlungu*) which is sub-standard, with poorly developed cobs, or partly infested with weevil or other forms of rot (*abolile*). A third category, *isiswenye*, consists of a quantity of the very best quality cobs from which the next season's seed is taken. So if you ask someone how much *maize* was harvested, the other two categories (*umdlungu* and *isiswenye*) are excluded from his/her answer. If fact, they may well also be excluded from the granary and certainly (later) from the tank, with only proper maize being stored and the *umdlungu* kept in an untidy pile in a hut for immediate consumption, beer brewing, or feeding livestock (or all three). The *isiswenye* is hung from the rafters of one of the huts till it is required in the following season. This does not mean that all the cobs designated as *isiswenye* (often estimated to be as much as 80 kg. of grain) will form the basis of the next crop, but only that the seed (*imbewu*) used will be taken from it, with what remains being eaten. Often, however, all three — *umbona*, *umdlungu*, and *isiswenye* — are initially placed in the granary, each separated by palm leaves, old sacking or suchlike. People with two granaries may use one for good maize, one for the sub-standard *umdlungu*.

The matter is made even more complex by the fact that the nature of the grain called *umdlungu* varies in relation to the size and quality of the harvest. If the harvest is large the cobs that are badly infested with weevil and rotten may be left in the garden for livestock to eat. The maize regarded as *umdlungu* under such circumstances might be of a quality not that far removed from the good maize. If the harvest is small however, very poor quality cobs may nevertheless be taken out of the fields and used, and the difference between 'good' and 'bad' maize will be more apparent. The *umdlungu* we were shown at one homestead was not all that different from the good maize, and the homestead head informed us that the 'real' *umdlungu* had been left in the garden for the pigs and cattle. In another case we were told that no *umdlungu* at all had been harvested, which reflected the fact that the harvest was of a good size and quality, with cobs not fit for human consumption being left in the lands for foraging livestock. In a third case, however, we were shown maize that we assumed was *umdlungu,* because of its poor quality, but we were told by the homestead head that it was not, because (he claimed rather inaccurately) the cobs were not rotten or weevil-infested.

6. Yields vary from year to year and a good measure of yields would have to take place over a period of around five to ten years to take into account both good and lean seasons. All three seasons covered in the course of this study were regarded as relatively good, and even here it was possible to record dramatic differences in the same garden from year to year. Time of planting is crucial. In 1997 those who planted in November were able to take advantage of early rains and appeared to do well, but those who planted in December complained that there had not been enough rain, which accounted for their poor crops. A number of households that regularly produced good harvests complained, in February 1998, that their maize plants had not developed well at all. Some of these had replanted in January, but this measure had not been very successful.

It is possible to get around some (but not all) of these problems of measurement. Since granaries are either roughly cylindrical or rectangular in shape, it is possible to measure their volume. If one has an idea of the volume occupied by the harvest of maize on the cob, and a specific portion of the same maize is removed from the cob and weighed, it is then possible to calculate the amount of maize in a granary in terms of kilograms of grain. If one also has a measurement of the amount of land used to produce this maize, a yield per hectare can be given. This does not, of course, solve the problem of inter-cropping mentioned above, nor tell one how much maize was consumed prior to the harvest, but it does give an idea of how much maize per hectare was harvested and stored.

The procedure used in this study was as follows: A wooden box with a volume of exactly one eighth of a cubic metre was constructed — internal dimensions of 500 mm. x 500 mm. x 500 mm. — and fitted with sturdy wooden handles. Four homesteads, representing people regarded as poor, average, above average and good farmers were visited and (with the permission of the head) careful measurements were made of the space occupied by maize stored in the granaries. The box was then filled with maize from the granary and weighed. Finally, the kernels were removed from the cobs, and both cobs and kernels were weighed separately. This procedure was followed for both maize and *umdlungu*. An average weight of the amount of grain represented by a cubic metre of maize on the cob, for both classes of grain, was thus obtained. An additional twelve homesteads were then visited, again representing a range of types of farmers — from homesteads headed by widows with little labour to those with many labour and other resources, and the contents of granaries measured to provide the amount of maize harvested and stored. The average weight per cubic metre as arrived at in the initial four homesteads was then applied, separately for good maize and *umdlungu*, so that we obtained the size of overall maize yields in kilograms for all sixteen homesteads.

Measurements were also made of the sixteen gardens used by each of these homesteads, using a metering device (a planimeter). Since this took place shortly after the harvest and the maize stalks were still in evidence, it was possible to identify and measure uncultivated parts and deduct this from the overall garden size so as to get a truer picture of the number of square metres under cultivation. The main difficulty is that many gardens are not strictly rectangular but (usually) merely quadrilateral, making it important to know the internal angles at which the sides joined each other.[6] Since we did not have any means of measuring the angles, recourse was later made to aerial photographs of the gardens in question, and the gardens plotted onto squared paper in accordance with their shape on the photograph and the measurements made in the field. A measurement in terms of square metres could thus be arrived at to facilitate expression of yields in terms of kilograms per hectare. This figure can be adjusted to take into account maize consumed prior to the harvest, based on the estimates of household members of how frequently they used such maize and in what quantities.

For example:
HG's granary is 2.6m in diameter and was filled with maize cobs to a height of 1.55 metres. The volume of maize in the granary is thus 8.232 cubic metres. Using the box mentioned above, one eighth of a cubic metre of granary maize yielded 53.2 kilograms of grain. The granary thus represented 3503.5 kilograms of grain. He also had *umdlungu*, which comprised 310.4 kilograms of grain, so his total harvest was 3813.9 kilograms. His garden is 13,982 square metres, or 1.3982 hectares, giving a yield of 2,727 kg. of maize per hectare.

It was found that the sixteen homesteads sampled had produced and stored a total of 22,614.3 kilograms of maize in the 1996/97 season. The average was 1,413.39 kg. per homestead. It has been estimated that the basic subsistence requirement for maize in Eastern Cape households is 175 kg. per person per year or 1,050 kg. per homestead of six people (ARDRI 1989, 30). On this basis, eleven of the sixteen homesteads listed in the table produced sufficient to meet their maize requirements in this particular season. The yields ranged in size from a low of 330.6 kg. to a high of 3,730.8 kg. Gardens ranged in size from 1,347 square metres to 20,225 square metres, and the yield per hectare ranged from 747 kg. (9.3 bags) to 4,683 kg. (58.5 bags). The average yield per hectare was 2,307.25 kg. (28.8 bags). The full results are detailed in the table below. It is worth

6 Some idea of the irregularity of field shapes can be obtained from the sketches in Hunter's *Reaction to Conquest* (1936), particularly the first of the two she gives — 'Sketch Map of Fields in a River Valley', between pp. 72 and 73.

pointing out that the two households with the lowest gross yields, when surveyed earlier and asked how much they had harvested, had said 'nothing'.

If one adds green maize to the average yield of 1413.39 kg. of maize per homestead, on the basis of 45 days consumption at 1 kg. per day, the figure rises to 1458.39 kg. (3215 lbs., 1.435 tons, 1.458 tonnes, or 18.2 bags of 80 kg.). The equivalent yield per hectare is similarly raised to 2352.25 kg. (approx. 29 bags). The figure of 45 days is a very conservative one. Hunter (1936, 84) states that the consumption of green food from the fields lasted for 'nearly four months' before the harvest of dried maize. Following Lipton (1977), if one quarter of the maize yield is consumed in this way then the harvest of homesteads that produce enough maize for the year represents three quarters of the total crop. Again, this is conservative; Lipton uses a figure of one third. For homesteads, which do

Table 1
Measured maize yields

Homestead	Area cultivated*	Yield in kgs.	Kgs. per hectare
MY	1346	632.0	4698
MS	7964	3730.8	4683
NQ	5300	2402.2	4532
NA	5458	1735.9	3180
HG	3982	3813.9	2727
MG	3742	881.1	2354
NT	6091	1374.9	2257
SK	3912	818.6	2092
NL	5614	1053.4	1875
NW	4212	744.2	1767
MN	2852	393.5	1379
NF	8891	1194.2	1343
GD	20225	2560.9	1266
NK	3544	437.0	1248
MP	4358	330.6	758
GL	6743	510.7	757

* This is the size of the area planted with maize. The size of the gardens themselves was sometimes considerably larger than this.

not produce enough maize for the year, the green maize eaten directly from the lands represents *more* than one quarter of the total crop.

The figure of 1 kg. green maize consumed per day is probably also much too low. Cairns and Lea (1990) reported that in the Nkandla district of KwaZulu 5.2 kg. of green maize (wet mass including cob and sheath) was harvested per day, six days a week, for 3 months prior to the harvest. Most homesteads ate 25% of the total harvest green, and this amounted to 190 kg. of the average harvest of 756 kg. Using the figure of 25%, it is likely that the average figure of 1413.39 kg. of maize per homestead for the sixteen Shixini homesteads sampled represents 75% of the total maize yield, putting the total average yield at 1884.39 kg. This gives some idea of the overall significance of cultivation in Shixini, though of course one would need to add in the substantial quantities of other vegetables which are produced, as well as the off take from livestock, for which figures are not available.

Even the conservative, average figure of 2352.25 kg. of maize per hectare is far higher than any previous estimates of maize yields in the Transkei. For example, a survey by ARDRI (The University of Fort Hare's Agricultural and Rural Development Research Institute) of the area served by the Lima Community Development Association, basically the coastal half of Xorha district, immediately adjacent to Gatyana, found that in 1987-88, a bad year (defined as being too wet), only 4% of households harvested more than 8 bags (640 kg.) of maize. Fifty-two percent harvested 'nothing', 25% got from 1 to 3 bags, and 18% from 4 to 8 bags. The average yield per hectare was 112 kg., and the long-term average was 315 kg. per hectare. So the survey concluded that 'in an average year the Lima area as a whole produces 30% of its subsistence grain requirements' (ARDRI 1989, 30). These low yields were obtained, they reported, despite a relatively high use of fertiliser, manure, cultivators and planters.

The ARDRI figures are close to those used by the Tomlinson Commission for rural black agriculture as a whole, namely 2,5 bags (500 lbs.) per morgen (about 280 kg. per hectare), which they compared unfavourably with the yield from commercial (white) agriculture of nearly 7 bags per morgen in 1959-60 (Lipton 1977). Another survey in 10 areas drawn from three districts in Transkei (Umtata, Tsolo and Centane) produced very similar figures to the ARDRI survey. It found that the average yield was 100 kg. per acre (247 kg. per hectare) and that in good years 30% of households grew enough food to feed themselves (Leeuwenberg 1977). Bembridge (1991), reporting on a survey in two Transkeian districts, Libode and Tsolo, puts average maize yields at 414 kg. per hectare, with the highest yields recorded being 2420 kg. per ha. In these districts, according to the survey results, 70% of homesteads did not meet annual grain requirements of 175 kg. per person. Similar figures for maize yields are provided by other commentators (e.g. Westcott 1977, Moll 1984, Phillips-Howard 1995).[7]

Other figures are slightly higher. Hunter (1936, 94) records that yields of 5.5 bags per acre (13.75 bags per hectare) were obtained from demonstration fields in Pondoland in 1929-30, with ordinary people harvesting only about 2 bags per acre (5 bags per hectare) from fields which were, on average, 7903 square yards (approx. 0.67 hectare) in extent (*ibid.,* 72-73). Assuming that these were 200 lb. bags, this equates to approximately 1250 kg. per hectare for the demonstration fields, and 455 kg. per hectare for ordinary people. A somewhat higher figure is provided by Mpambani (1994, 13), where Transkei maize yields are estimated at 624 kg. per hectare (i.e. 7 to 8 bags of 80 kg. each). Significantly, the only case of a survey using careful measurements of crops grown in the Transkei is one done in Amadiba location in coastal Bizana district in 1985, which recorded average yields (of all crops) of 25 bags per annum, high enough for basic staple requirements (see Beinart 1992, 185). It does seem that land holdings were unusually large there by Transkei standards (six hectares per household), so yields per hectare may still have been relatively low.[8]

---

7 Beinart (1992, 184) gives much higher figures for Tsolo, but this is clearly at variance with the figures provided in the published version of the conference paper by Westcott (1977) that he cites.

8 Institute for Management and Development Studies: Socio-economic Survey of Area 6, Eastern Pondoland (Bizana Sugar Project). Umtata, University of Transkei, 1986, cited in Beinart (1992). Beinart does not cite separate figures for maize yields from this survey, and I have not been able to access the original.

In stark contrast to most Transkei survey findings are those obtained by economist Norman Bromberger in a study of a peri-urban area of KwaZulu-Natal. Here, with careful measurement and sampling of maize plants prior to harvest, Bromberger found mean maize yields from small gardens to be 4200 kg. per hectare (University of Natal 1984; Bromberger 1998, pers. comm.). This was an area fairly well endowed with rainfall of 1000 mm. per annum, similar to Shixini's rainfall figure, and where people were relatively well off and able to afford regular inputs such as commercial seed and fertilizer. No doubt the small size of the cultivated areas (gardens were from 0,1 ha to 0,3 ha in extent) made intensive labour possible and also contributed to the high yields. The method used by Bromberger was to count the maize cobs on the stalks in sample strips, then to collect cobs that had survived through to harvest stage and weigh the grain. The average mass of grain from the cobs was then applied to the number of cobs counted earlier and then extrapolated to the whole garden. A difficulty with this method in Shixini would be that people select out the smaller, underdeveloped and diseased cobs during the green maize stage, and leave the better ones to reach full maturity, so one could not assume that all cobs counted early on would be similar in size at maturity.[9] More recent findings based on PRA surveys in KwaZulu-Natal have estimated that food grown at home accounts for between 25% and 45% of rural household food consumption. In addition to this, however, between 25% and 60% of household food was obtained from other households in exchange for labour (May 1996). This differs markedly from the finding from the large SALDRU survey conducted in 1993, which estimated agricultural production in rural households in Natal at only 6.8% of total household income (including maize, milk, eggs and livestock products).

Clearly, survey methods produce widely contrasting findings yet the dominant trend is for surveys consistently to underestimate the role of agriculture in former homeland areas. Yields such as 300 kg. per hectare in an average year are, as Lipton (1977) pointed out, simply too low to

---

9 Another difficulty would be in selecting a 'typical' sample area within the garden, given the uneven distribution of maize plants linked to the practice of broadcasting seed and to intercropping, described earlier.

believe given comparative statistics from other parts of Africa and Southern Africa. Yields among black subsistence farmers in the 1950s in Lesotho were between 7 and 8 bags per morgen, and between 5 and 9,5 bags per morgen in Swaziland (*ibid.*). Equally clearly, studies based on careful measurements of yields (rather than on responses to a survey questionnaire) show that there is a lot more maize being produced in places like coastal Transkei and rural KwaZulu-Natal than is conventionally accepted. Those who have pondered the results of questionnaire-surveys in this respect have sometimes asked why rural people bother to cultivate at all, speculating that perhaps it is to maintain claims to land (which of course begs another, similar question). The answer is evidently that they continue to cultivate because they find it worthwhile — a rational choice made on economic grounds. In Shixini, people do not practise agriculture simply because they identify as rural Xhosa, though this may be important too, and being part of a mutually supportive community that values its rural Xhosa identity and regards itself as living a 'traditional' lifestyle is a vital ingredient in facilitating agricultural production, as we shall see below.

It is possible to come to the conclusion then, that agriculture in the Transkei is possibly not, as almost all commentators assume, 'hopelessly unproductive', and that some Transkei farmers, in places like Shixini, know what they are doing and are skilled enough to produce maize yields which in good years are comparable with the average yields of dry-land commercial farmers in South Africa's 'maize triangle'. In poorer land in the Western Transvaal farmers get around 2000 kg. of maize per hectare (Westcott 1977), and Bembridge (1990, 21) gives the overall maize yield from commercial farmers as 2100 kg. per hectare (which he compares with the yields from South Africa's less developed areas of 0.78 tonnes per ha). In the highly mechanised maize production schemes established in former Transkei by TRACOR in the 1980s, yields were around 20 bags (1600 kg.) per hectare, with up to 40 bags (3200 kg.) per hectare in 'exceptional' years (Cloete 1985). What is remarkable about some Shixini farmers is that they are able to achieve good yields despite a complete absence of state support and the severe constraints imposed by shortages of labour and implements, and without high input costs linked to mechanization, chemical fertilisers, insecticides and commercial seed. This does not mean that they do not want these sorts of inputs — on the contrary, they want them badly, but they are simply not available.

In asking to what extent one can generalise these findings, it is necessary to point out that even in a single local area there can be important variations between local communities with regard to the extent to which they are involved in agrarian activities. For example, in the Dwesa-Cwebe area, which straddles the Mbashe river, the boundary between Gatyana and the adjoining district of Xhora, some communities have more or less given up agriculture altogether while other adjacent communities are still firmly committed to it. On the Dwesa side of the Mbashe (in Gatyana) live the present day descendants of Mfengu people who were settled there in the late nineteenth century as colonial allies, in an attempt to reduce the influence of the Xhosa under Sarhili. Today the members of this community practise very little agriculture, though their ancestors were commercially successful cultivators in the early 1900s. On the northern side of the Mbashe, in contrast, the more conservative, primarily Bomvana people are still heavily engaged in subsistence agriculture (Fay and Palmer 2000, 202-3).

These sorts of local variations aside, Shixini is probably typical of the communal areas in the coastal part of Gatyana district, and of the Transkei coastal belt as a whole. Most of this area has not been subject to agricultural betterment, enjoys good rainfall, and retains a degree of social and political stability. The Bizana survey mentioned above bears this out, and some of the figures from KwaZulu-Natal show that a reasonable standard of agricultural production is probably typical of many parts of that region also, at least where rural communities still have access to land and have not been destroyed by agricultural 'betterment' schemes. However, there is no question that many parts of Transkei, subjected to betterment schemes, with higher population density, a high degree of landlessness, and with lower rainfall, are not likely to produce average maize yields of the same size as Shixini or other higher rainfall coastal areas where betterment has been kept at bay and most homesteads still have land. Nevertheless, survey results in general undoubtedly underestimate agricultural production, and few take into account the whole range of crops and other activities that contribute to subsistence (see Beinart 1992; Shackelton et al. 1999).

The large variations in yields per hectare in Shixini are understandable. Inputs (especially labour) vary considerably across households, and when all other things are equal maize yields depend heavily on the amount of labour devoted to weeding. All other things are not equal, of course. Heron has shown how important it is to plough and plant at the right time in Shixini, and how few days there are available to do so given the vagaries of weather conditions and shortages of oxen, ploughs and labour (Heron 1990). Kuckertz (1990, 189) has made a similar observation. Then there are differences in the type and amount of fertiliser and seed used, the depth at which ploughing occurs, the fertility of different types of garden soil, etc. And in the final analysis, although farmers try to control as many of these variables as possible, there is an element of unexplained luck, good or bad. Farmers sometimes simply cannot explain why a normally productive garden suddenly produces very little maize, or why the maize seeds hardly germinate one year and another planting is necessary, or why a garden planted in early February, instead of at the optimum time in November, produces a bumper harvest. The differences in yields between the 16 households in the above table are also linked to differences such as household size and livestock holdings. Heron (1990) and others (see Beinart 1992) have shown that livestock ownership correlates with maize yields, and it should be obvious that smaller households with little labour are likely to farm smaller areas less efficiently than others. They do not correlate with the gender of the household head — seven of the sixteen households listed have female heads, and these are by no means the seven with the lowest yields or the smallest gardens.

The implications for rural development

Certain policy implications are implicit in the above which need to be considered in view of new policy directions recently signalled by the Departments of Agriculture and Land Affairs and the criticisms that have been levelled at these policy statements from a number of quarters. Perhaps the starting point here is the characterisation, by the Department, of rural people in the former 'homelands' as the 'poorest of the poor', and the perception that such people are agriculturally unproductive, beyond providing a little food for household use. The Shixini case, together with some of the comparative material outlined, are a reminder that rural areas are not homogenous and point to a need to question the conventional perception of rural Transkei agriculture, and by extension agriculture in all of the former 'homelands', as being inefficient and in decline. As Merle Lipton (1977) pointed out, the notion that yields in the Bantustans were very low derives primarily from the Tomlinson Commission, which also *over*estimated maize consumption. The large gap between yields and needs could conveniently be used to argue that it was necessary for black rural dwellers to engage in wage labour in the mining and industrial sector. We need to beware of perpetuating this image of Transkei agriculture, and of the new vested interests that it may serve.

Currently, the new proposals by the National Departments of Agriculture and Land Affairs are one of two major policy initiatives designed to provide economic development in black rural areas that were formerly part of the 'homelands'. The other is that associated with the Spatial Development Initiatives, of which there are several in the country as a whole. We shall first consider the development implications of the Wild Coast Spatial Development Initiative (WCSDI), since it has been in existence for about four years already, before turning back to the question of commercial agriculture.

The WCSDI is the most significant recent development initiative for the coastal belt of Transkei and one of two SDIs in the country that have been billed as 'agri-tourism' SDIs, started in 1996 by the Department of Trade and Industry. The SDIs are designed to develop the economic potential in some of the country's undeveloped areas and, at the same time, to benefit the poor residents of those areas. They are linked to and part of GEAR, the state's macro-economic policy framework designed to encourage economic growth in South Africa as a whole (Kepe 2000). The progress and potential of the WCSDI needs to be assessed in relation to the findings discussed above and the current debates about agrarian development in rural South Africa, because, in theory at least, the SDIs are designed to provide for the active participation of local rural communities, and to develop commercial and employment opportunities for local people (Mahlathi, 2000, 250).

The WCSDI, in the words of its Director, is not a "comprehensive development strategy", but attempts to facilitate investment by both government and the private sector in six "development nodes" along the 300 km long Wild Coast. These nodes include commercial agricultural developments established during the 'homeland' era but neglected since the transition to a new regional government system, such as the Magwa tea estate, and tourism sites and facilities, such as those in the vicinity of existing nature reserves. One of the latter, the Dwesa-Cwebe nature reserve, is about 40 km by road from Shixini, the Dwesa section of the reserve being in the same district, Gatyana.

The basic elements of the WCSDI, in terms of the development of the identified 'nodes' are, firstly, 'public-private sector partnerships' in which government provides funding for infrastructural upgrading and development in the selected 'nodes', and where this is accompanied by private investment in projects such as the upgrading of tourism facilities and the establishment of craft centres. The second major element is 'community partnership' in the sense of the investment taking place in consultation with local communities, the establishment of employment opportunities for local people, and commercial development which has income-providing spin-offs for locals, such as craft centres and the use of land for agro-industries. Theoretically, development efforts that focus on agro-industries and non-farm activities could stimulate agriculture by making income available for investment in livestock and cultivation (Meagher 1998). However, mundane goals such as food self-sufficiency for local communities do not feature as part of the SDI plans. On the contrary, the bias against household food production, evident in the new national policy directions discussed below, are evident also at the level of SDIs. The Director of the WCSDI has, for example, has indicated that there is a need to make "a transition away from inefficient and low-productivity agriculture" (Mahlathi 2000, 251).

While some progress has been made in the case of the WCSDI, it has been slow, and beset with a variety of problems. Government has made some funding available for infrastructural development and upgrading, and some job creation has taken place as a result, but a great deal more is needed. The very poor state of the roads providing access to the Wild Coast is a major problem, a disincentive to tourism and to investors. Much of the state's spending so far has gone into road maintenance and upgrading (Fay and Palmer, 2000; Daily Dispatch 2000, 2000a). Attention has also to be paid to electricity supply, water provision, and sewage disposal.

Land reform in the identified development nodes is also a difficulty (Kepe 2000; Fay and Palmer 2000). Private investors are reluctant to commit funds until the question of tenure is settled, and in this respect the progress has been slow. This is partly due to divergent and conflicting interests in the affected communities, and to more general communication problems making it difficult to secure community participation in proposed developments. There are also weaknesses at local government level, where there is a lack of capacity, and which is currently in the throes of re-organisation (Mahlathi 2000; Fay and Palmer 2000). The rural district and regional councils, and the transitional representative councils, are new and still finding their feet. There is still uncertainty over the role of the traditional authority figures (the chiefs and headmen, including their roles as heads of the Tribal Authorities of the previous era), and also about the proposed new structures for local government in which the urban municipalities are due to take on responsibilities for the entire districts in which they are located. The financial and organisational implications of this are enormous.[10] At the level of provincial and national government there have been difficulties of co-ordination due to different demands and requirements across different departments and the absence of an integrated legislative and regulatory framework (Kepe 2000;

---

[10] I am grateful to Dr Etienne Nel of Rhodes University, Grahamstown, for drawing my attention to this.

Mahlathi 2000). In short, the 'enabling environment' envisaged by government to attract private investors and facilitate partnership with local communities has not yet been created. Some of the projects that have been announced in the past have gone ahead, although often in scaled down form, and with some benefits to local people, but others have been abandoned (Palmer et al. 2000). But even if the WCSDI were to overcome the above-mentioned difficulties and proceed as hoped, it is likely that the benefits to the 'rural poor' will be limited to relatively small numbers of people from communities immediately adjacent to the development 'nodes' (Kepe 2000; Palmer et al. 2000; Fay and Palmer 2000), and that the impact on job creation and poverty alleviation will be small.

What this points to is the continued importance, in the foreseeable future, of land-based livelihoods for the majority of Wild Coast residents. In the light of this likelihood we need to return to the recent policy announcements from the national Ministry of Agriculture and Land Affairs, mentioned in the introductory chapter. In this I draw primarily on the policy statements published by the Ministry, examining the new policy direction in terms of the Shixini data. The critique also draws on the work of Professor Ben Cousins and some of his associates at the Programme for Land and Agrarian Studies (PLAAS), at the University of the Western Cape. Cousins is a respected commentator on agrarian issues in South Africa, and PLAAS has been at the forefront of research into agrarian issues and policy over the past few years.

The new agricultural development and land reform policy has emerged in a number of policy announcements from September 1999 onwards (Cousins 2000), culminating in mid-2000 with the final draft of what is titled the 'Integrated Programme of Land Redistribution and Agricultural Development in South Africa' (NDALA 2000a). A number of the provisions made are welcome, according to commentators, in the sense that they recognise past failings (e.g. in securing land for redistribution) and attempt to remedy them, (Cousins *ibid.;* Lahiff 2000). Other provisions of the new policy however, are extremely worrying to many of those engaged with agrarian reform in South Africa. A major criticism is the emphasis on the provision of support (including land reform) aimed at the development of 'commercial' farmers, and the relative lack of concern with the millions of rural dwellers who fall into the category of 'subsistence' farmers or would-be 'subsistence' farmers – i.e. those who produce primarily for household food consumption, or who would probably do so if only they had access to some arable land. The 'final draft' of the new policy certainly mentions the need to alleviate rural poverty, improve nutrition and assist with the provision of household food security among the rural poor, and claims that the new policy is aimed at those who want to farm 'on any scale'(*ibid.,* 5).[11] At a later point it states that: 'Most beneficiaries seeking the minimum grant will use it to establish small farms and food gardens for home consumption to supplement nutrition *(ibid., 9)*.' However, the provisions made to facilitate such aims are woefully lacking or inadequate, beneficiaries of this kind are subjected to the same kinds of constraints as the larger, commercial farmers, and the implication is that this category of households is also expected to farm commercially (although there is ambiguity and lack of clarity on this in the document).

The most recent policy document pays considerably more attention to the class of smaller grant applicants that the earlier policy statements, and seems to indicate a concern, at least in the statements made, for the non-commercial, 'subsistence' sector. This is a departure from the earlier documents, where it is the class of 'commercial' farmers who are seen as making a contribution to economic development, and who are targeted as the beneficiaries of state support and land reform. [12] Yet the economic contribution of the 'subsistence' sector is not recognised, apart from the fact that such households may provide a 'food safety net' (see NDALA, 2000). This term is still used to describe those who will farm for household consumption, and the implication is that there is still a clear dichotomy between 'commercial' agriculture and agriculture which provides

---

11 Page references to this policy document are based on the version published on the internet; see http://www.nda.agric.za/

12 In commenting on the April 2000 policy document, the Nkuzi Development Association pointed out that "references to the rural poor have almost disappeared for policy pronouncements of late, which gives grounds for serious concern." (Nkuzi 2000, 2)

goods primarily for household consumption. The former, by implication, is valued and economically useful, the latter is not.

The overall tenor of the new policy is towards support for 'commercial' farmers. Grants of between R20 000 and R100 000 are to be made available to acquire land and/or to undertake farming operations, and the procedures for the implementation of the scheme are clearly and explicitly geared towards the establishment of commercial ventures. Every grantee has to provide an 'own contribution' in cash or in kind, which in the case of the smaller grants amounts to a minimum of R5 000 and could be in the form of labour, capital assets (e.g. equipment) or land. This is not too high, it is claimed, to exclude the rural poor, but this is highly questionable (Lahiff 2000). Larger grantees are seen as probably having to secure bank loans, using standard banking procedures, to provide the 'own contribution' of up to 50% of the grants they receive. They will have the expertise required to undertake a commercial farming venture and will draft a 'farm plan' (including budget) that will be submitted to a local agricultural officer for an opinion. Only then will the plan go to a provincial grant approval committee for consideration. Successful applicants (at all levels, it seems) will have to participate in 'training courses and activities designed to assist them in successful operation of their farms and gardens' (*ibid.,* 10).

The households applying for larger grants will obviously already have a relatively good income and level of education, and they are envisaged as employers of agricultural labour. This is a segment of the rural population already relatively better of than the majority. Cousins (2000, 4) estimates that even the most optimistic of estimates would not make the potential number of such farmers greater than 20,000 or 30,000 in the country as a whole. Even taking the recipients of the smallest grants into account, millions of rural households will not benefit from this scheme, except as possible employees of the 'commercial' farmers, and they are more or less ignored for development and land reform purposes. The ministry, however, suggests that there could be around 200,000 applicants for smaller grants, and some 30,000 for larger grants.

The 'commercial' orientation of the new policy is seen also in the land reform measures associated with the grants scheme. In all cases, large and small, the policy requires security of tenure in the form of title to the land. In the case of larger individual grantees seeking to acquire land that is for sale or making application for state land, there is little difficulty here. These will proceed in the normal manner, with land transfer being effected and title registered. Similarly, groups of beneficiaries pooling their grants and being granted land by the state or purchasing land for sale may do so in terms of sectional title. The grant that they receive may be used partly to identify land for sale (through an estate agent or a 'design agent'), purchase land, pay for transfer costs, etc. Where the new policy does raise considerable fears is in recommending that communal land, too, may be purchased under the scheme, either individually or as a group *(ibid.,* 11). Individual or sectional title to the land is seen as increasing the 'efficacy of land use' *(ibid.,* 13). Communal land could be offered for sale to members of communities on that land, with 'communities and traditional leaders' responding to requests for sectional title. In such a case 'the traditional authority would be the seller ... and with the exchange of money, the title would be transferred to the name of the buyer' *(ibid.,* 13). The grant would be used partly to pay for the transfer. In such cases the seller (sometimes referred to as the 'community', sometimes as the 'traditional authority') and the buyer (recipient of a grant and member of the community) would enter into the agreement voluntarily. However, since a requirement of the programme is that the land to be used should be marketable 'in order to retain its value' *(ibid.,* 4) this means, in effect, that people residing on communal land would not be eligible for a grant unless they were able to purchase land from the community (i.e. pay for land that they have a right to use anyway), and that such land could be sold to non-community members, otherwise it could not be regarded as 'marketable'.

The problems associated with these rather vague and naïve proposals are many, as a long-standing debate over the difficulty and desirability of introducing freehold into South Africa's communal areas indicates (Cross and Haines 1988; McAllister 1992). The policy-makers show no awareness of this debate or of the potentially disastrous consequences of introducing individual title into communal areas, such as the creation of landed and landless segments through poor people selling their land to those able to buy it, rifts between traditional leaders and their people, corruption, and increased poverty. Instead, it is claimed, rather astonishingly, that under the provisions made 'the general character of communal property would not be altered fundamentally' *(ibid.,* 13). If one looks at the case studies in chapter two in terms of the new policy provisions, some disturbing questions arise here, particularly if one considers households that are temporarily unable to cultivate. The case studies demonstrate that Shixini households in desperate situations can temporarily cease to practice agriculture, but remain part of the social and moral community through participation in agricultural activities of others, until they regain the ability to farm for themselves once again. The new policy seems to close off such opportunities and prevent this from taking place, condemning such people to perpetual poverty.

The proposal that the money received for the land so bought should be used to develop the community through building schools, clinics, and improving infrastructure, is similarly naïve, and no mechanisms are proposed to facilitate this.

In a recent comment on the new policy direction, Cousins questions the current ability of the state to undertake meaningful land reform due to problems within the Departments of Agriculture and Land Affairs such as policy confusion, weak leadership, poor morale and the resignation of many key personnel since the new Minister took office. Land rights in the former homelands are in a state of uncertainty, and the new commercial farmer policy, he feels, 'attempts to smuggle into policy a mechanism to individualise communal tenure systems' (2000b). There are also fears that state land in the former homelands held under communal tenure may be transferred to tribal groups under chiefs, in place of the previous Communal Property Associations, which presented a potentially viable and acceptable means of formalising and securing tenure for large numbers of rural dwellers (Cousins 2000a). The new policy makes mention of communal property associations as an alternative to sectional title, but provides no evaluation of the CPA instituted under the previous Minister of

Land Affairs, and no indication of how this would meet the needs of land marketability, seen as so essential for the scheme to work

There are also major difficulties with regard to the implementation of the proposed new policy. The approval process for the grants scheme is to be moved to the provincial level, as already indicated, and a vital role is seen for local agricultural officers, who must scrutinise and give an opinion on the feasibility of farm plans, 'including its agricultural potential, value of the land relative to market prices for that of comparable quality and access to water, cash-flow projection, and environmental assessment.' They also have to ensure that the seller is the valid title-holder to the land in question, and give a view on the overall feasibility of the proposal. This is a lot to ask of even the most well trained agricultural officers anywhere, and it is doubtful whether officials with this level of training and expertise exist at provincial level in South Africa. Cousins (2000b) is rightly sceptical of the ability of the provincial departments of agriculture and land affairs to implement the commercial farmer programme, due to the fact that they are short on capacity, funding and expertise and, like their equivalents at the national level, suffering from low morale. In 1999 extension officials of the Eastern Cape's department of agriculture were unable to undertake any extension work at all, due to a shortage of funds. Nearly 90% of rural people in the former homelands of South Africa have had no contact with agricultural extension staff, according to a survey conducted in 1996 (*ibid.*).

These are unfortunate developments from the point of view of those who champion the cause of the many poor rural residents of the former homelands, and a move away from earlier policies that seemed to have the interests of such people at heart. In a policy document released in 1998 under the former minister, for example, it is pointed out that nearly three-quarters of the 16 million South Africans who live in poverty are to be found in rural areas, and that many of these live in a state of 'food insecurity' (defined as 'a lack of access to adequate, safe and nutritious food') (NDALA 1998, 6). Policy at that time included increasing the production and improving the food security of those who practised agriculture in these areas, and it was pointed out that 'it is necessary to understand how people in rural areas create livelihoods.' (*ibid.,* 7). Unfortunately, very little was done to realise these aims, a situation that now seems likely to continue. The new policy has failed to address them, and it has failed to take on board many of the lessons learned by those involved in agrarian and land reform issues. Instead, it looks depressingly like the kinds of policies that were prevalent in the era of the 'independent' Bantustans (see McAllister 1992).

The current policy direction is misguided, it is argued, because even if it is successful, commercialisation of the type being promoted will not provide the level of development required to benefit the masses of rural poor. It will require high inputs in terms of finance, expertise, management, organisation and manpower, to the neglect of the millions who are struggling to eke out an existence in rural areas. Commercialisation through attempts to establish a class of small to medium scale farmers has failed in other parts of Africa, and earlier attempts at commercialisation of the kind currently being envisaged in South African black rural areas have, with a few exceptions, not succeeded.[13] The general idea of creating a farming 'class' in areas like the Eastern Cape has been around for some time. It found earlier expression in a variety of forms (e.g. the 'commercial farmer' programme associated with the Development Bank of South Africa, a decade ago) and was justly criticised by many (e.g. Beinart 1992; Deliwe 1997).

While commercial agriculture is certainly required and should not be neglected, it is not an appropriate development path for the alleviation of poverty in the communal areas. A more equitable and viable alternative, it is suggested, is to refocus on the large numbers of small scale 'subsistence' producers, those producing at least a part of their subsistence needs from the land, and to develop and expand this sector. The arguments in favour of such a strategy have been made by a number of writers.[14] Beinart (1992) pointed out that Transkei households tended to utilise a

[13] There is ample evidence for this in relation to the Eastern Cape – see, for example, Cloete 1985, Deliwe 1997; Porter and Phillips Howard 1997; McAllister 1992).

[14] See, for example, Porter and Phillips-Howard 1997; Ottaway 1996; Beinart 1992; Shackleton et al. 2000; Adams et al. 2000; Cousins 2000)

wide range of strategies to secure a livelihood, and that they were likely to continue to do so. With current high levels of unemployment and continuing mass retrenchments in the mining and other industries, however, access to land and homestead food production is assuming greater importance once more for those who have lost access to remittances (Porter and Phillips-Howard 1997), and any strategies which ignore such homesteads or, worse still, deprive them of access to land and other natural resources, must be questionable on political, economic and moral grounds (see Ottaway 1996).

In contrast to other parts of Africa, including South Africa, occupational change, income diversification and a move towards non-agrarian activities are not significant trends. Migrant labour and most of the other means of earning cash have existed in Shixini for generations, and there is little evidence for a rapid move to diverse means of earning a living, as witnessed in other parts of the continent (Bryceson 1998). On the contrary, there are fewer and fewer migrants in paid employment, with unemployment rates in the Wild Coast area being in the region of 73% (Mahlathi 2000, 251). What is relatively new are the sizable social pensions (R520 per month) being paid to people, although pension payments as such are not particularly new. For a few, this has made it possible to eke out a rural living without spending money or labour on agriculture; to a small number of others, it provides the capital for small-scale trading in liquor or groceries; but for many, it has provided a valuable supplement to homestead food production rather than a substitute for it, and a means of investing in agricultural activities. As we have seen, there is little scope for individual wage employment in Shixini or nearby. The main form of occupation in Shixini, and the main form of labour, is agrarian. A small proportion of the agricultural work is monetised (see below), but this is not new, and it is tied to agrarian activities and an agrarian rationale. Certainly, reliance on remittances and pensions means that we cannot characterise Shixini, or any other part of rural South Africa, as purely agrarian, but this has been true for well over a century. No doubt the wages paid to Shixini migrants have risen since South Africa's transformation process got under way in approximately 1990, but since many migrants work in enterprises such as commercial agriculture and fishing, these rises have probably not been very high in real terms, and growing unemployment is likely to mean more rather than less dependence on homestead food production in the near future. What this points to is the need for development styles that recognise the importance of homestead production, be it subsistence or surplus, and the relationships between homesteads that facilitate this. What is required is a revival of the peasant agriculture largely destroyed by apartheid (Ottaway 1996), which has the potential to provide livelihoods for millions.

More recently, it has been calculated that the economic value of 'land-based livelihoods' in South Africa's communal areas, though not recognised by the current policy direction, is in the region of R13.28 billion per annum, calculated in terms of the value of crop and livestock production, and of various forms of natural resource harvesting (Shackleton et al. 2000). Most of the value produced does not appear in the national accounts, but this sector does also have the potential for small-scale commercial production (e.g. production for local markets or for informal exchange in local areas), which should be encouraged (Lipton et al. 1996). Food self-sufficiency and commercial production should not be viewed as two different things – the latter can develop out of and contribute towards the former, with the right kinds of policies and encouragement. The material presented above in relation to Shixini farmers provides support for these views.

What kinds of policies and support are required? Here too there is a fair amount of consensus based on experience in other parts of Africa. What is required is the provision of services and infrastructure on which rural households can draw, voluntarily, to improve production. The provision of agricultural inputs is a basic need, sorely neglected at present. Extension services offering advice and information are also of basic importance, and also sorely lacking at the present. As indicated above, Shixini farmers are able to produce crops of substantial economic value, and this ability would improve with the ready availability of basic inputs and advice on which they could draw as required by their needs determined by themselves, not from the outside in terms of some form of involuntary and socially disruptive 'scheme'. The philosophy needs to be one where local knowledge, skills, and practices are recognised and built on, rather than denigrated and transformed. The development of local markets, storage facilities and transport would help to stimulate commercial production. That the capacity for it exists is demonstrated by the figures on maize yields, above (see also Lipton et al. 1996; Bernstein 1996).

That the demand for agricultural inputs exists is certain. Familiarity with many Shixini farmers and my own involvement in providing agricultural inputs in Shixini in the late-1980s (McAllister 1994) has demonstrated that there is a large and widespread demand for inputs such as seed, fertiliser, fencing materials, insecticides, and other goods in the area. These basic products are very difficult to obtain locally, requiring time and money for a trip to Willowvale town some 30 km away, and the expense of arranging transport to bring them back home. The implications of this are that, given the right kind of support, household-based agriculture in Shixini and similar areas would grow. This could foster homestead food self-sufficiency, the evolution of smallholder commercial agriculture, and income diversification around increased agricultural activity.

The avenues for such development need to be provided without embarking on a strategy that could undermine food production in a section of rural households. In this respect the material in the next chapter on *how* production takes place in Shixini (i.e. through co-operative work) is important, since it demonstrates the possible consequences of creating a division between farming and non-farming households. It shows that even households which do not cultivate are involved in the agricultural activities of others (e.g. as volunteers at a work-party); that those who farm depend on a sense of community and collaboration through which they receive assistance from, and give assistance to, others; that smaller and poorer households which produce only a small portion of their food requirements through agriculture would probably produce nothing if cut off from their networks of reciprocal assistance; and that networks of mutual assistance are an important means of maintaining homestead production and redistributing surpluses that occur. In short, creating a division between commercial 'smallholders' and others would in all likelihood undermine the ability of *both* types of households to produce. Looked at in another light these labour arrangements also illustrate the kinds of local institutional arrangements that make homestead production possible. This should not be taken as a plea for outside agencies to try and harness and manipulate them for 'development' purposes, merely that they be recognised as valuable social resources which, if disrupted and changed through ill-considered agricultural and land-tenure reforms, could have the opposite effect of what is intended.

Also required is a policy that makes more arable land available to the rural residents who are landless, expands the number of people producing on a small-scale basis, and provides security of tenure that, in the case of places like Shixini, strengthens individual rights within a system of communal title. Communal tenure is not, in itself, an impediment to agrarian development in Africa, as long as rights are secured and vested in the people concerned (see Cross and Haines 1988; de Wet 1988). As we have seen above, people in Shixini have land rights by virtue of their residence in the area, granted and recognised by the local territorial community (the sub-ward under a sub-headman), and formalised by the headman/chief and the Tribal Authority. Ultimately, therefore, prior to the advent of a democratic government in South Africa, land rights were vested in the chief and the TA, and this still seems to be the *de facto* situation, and one of the areas of jurisdiction currently being contested in negotiations between traditional authorities and government. Existing rights in areas like Shixini need to be secured, confirmed and legislatively formalised through a means acceptable to Shixini people, perhaps along the lines of an institution such as a Communal Property Association (CPA). Current practices for the allocation, use and inheritance of land work relatively well, in general, but they need to be given legal recognition, and rights need to be secured within an acceptable system of local government. Given the growing political awareness in Shixini, and the association of Tribal Authorities with the apartheid regime, it is important that land rights be secured and agrarian development take place in a manner acceptable to rural people, including the youth. In addition, as indicated above in the discussion on gardens and fields, Shixini people have shown that the way in which they use land is flexible and innovative, responding to environmental changes. The continued right of landholders to make decisions regarding the management of their land needs to be vested in new tenurial arrangements made in collaboration with the community as a whole. It is not just a matter of smallholder farmers having specific needs (Machete and Mollel 2000, 341), but different groups or communities of smallholder farmers have to be understood, since they may have different priorities, different ways of organising themselves, and may have made different adaptations to the environment. In addition, as is frequently pointed out, the position of female-headed households is of particular importance, with the land rights of women requiring special attention, and this too may vary from local area to local area.

In the past, as indicated above, the rights of Shixini people have been threatened through the partial implementation of 'betterment' schemes which required the approval of the chief and the Tribal Authority, and which was imposed on ordinary people without their consent. In the case of one sub-ward, their land rights were lost due to the development of a forestry project on their land, also with the approval of the chief. This does not mean that new tenurial arrangements should by-pass the chiefs altogether, simply that the role of the chiefs needs to be carefully defined and agreed on by those involved. It needs to be remembered that customary forms of tenure such as the one prevailing in Shixini were modified through measures such as the Bantu Authorities Act (1955) and the imposition of state control over communities through chiefs and headmen. Tenurial reform can potentially restore the rights of people in relation to their traditional leaders, and need not necessarily be seen as a threat to such leaders. Tenurial reform of the kind envisaged in the new agricultural development policy, however, is likely to fuel potential antagonisms between leaders and people, introduce a market in land in formerly communal areas, leading to the creation of a class of landowners and another of landless people, unable to partake in agricultural activities at all.

In Shixini, providing greater security of tenure would not only encourage more investment in agriculture, but also strengthen the social fabric and reinforce the existing morality of mutual help, and the networks of co-operation among farmers, thereby strengthening their ability to achieve food self-sufficiency and to produce commercially, over and above household food requirements. It is to these social and organisational factors that we now turn. Given the dependence of Shixini people on their land, the question that arises is: *How* do Shixini households manage to cultivate at all, given uncertain rainfall, poverty, lack of inputs and state support, a degree of insecurity of tenure, and shortages of labour and draught power? And how, to return to another theme of this book, is cultivation related to Xhosa identity? The answer to these two questions turns out to be the same. In their agricultural activities Shixini people depend, to a very large extent, on co-operation and mutual help between homesteads, which they see as a manifestation of Xhosa custom and tradition, an aspect of their rural Xhosa identity. In the following chapter the nature and extent of Xhosa work parties are considered against the background of a review of the literature on work parties in Africa, and the historical circumstances within which Xhosa work parties assumed an importance beyond which they had held previously. In subsequent chapters the question of how this is linked to beer drinks and to Xhosa rural identity are considered. In this I put the question of policy temporarily to one side, although the implications should be fairly obvious. I am concerned with unravelling the complex nature of the relationship between beer drinking and agricultural labour, and in an analysis of how this relates to the Shixini concern with 'building the homestead'. I return to policy issues briefly in the conclusion.

4

Co-operative labour in Africa: an overview

Co-operative, extra-familial labour in the form of 'work parties' or other forms of voluntary productive collaboration is historically widespread in Southern Africa and other parts of the continent[1]. Although only one form of labour among many, and sometimes insignificant, in many areas co-operative labour plays an important contemporary role, particularly in agriculture, and in recent years it has been the subject of some careful analyses by anthropologists.[2] It is necessary to review the existing material on African co-operative labour in order to place the Shixini material in comparative perspective and to illustrate how the analysis presented relates to and differs from earlier approaches. In this my concerns are eth-

1 Reviews of some of the evidence on work-parties in Africa may be found in Goodfellow (1939), Shaw (1974), Moore (1975), Sansom (1974), Saul (1983), Swindell (1985), Kuckertz (1985) and Painter (1986). Co-operative labour is only one of many forms of work on the continent, and is often less important than other forms, such as wage labour or household labour (Swindell 1985). Thus some of the work on African labour in the context of agricultural development and agrarian change makes hardly a reference to it (e.g. Berry 1993).

2 Donham (1985), Kuckertz (1985), Geschiere (1995), Worby (1995), and Englund (1999).

nographic and anthropological rather than directly related to questions of rural development. Readers who are primarily interested in the practicalities of how co-operative work takes place in Shixini might wish to skip this review and go directly to the second section of chapter five.

It is widely reported that co-operative work in Africa is concluded with the ceremonial or festive consumption of alcoholic drinks such as maize or millet beer and/or food, often described as taking the form 'beer-parties'.[3] Although this is an obviously important aspect of the work that precedes it, judging by the frequency with which it is mentioned, few writers have paid much attention to these events.[4] One aim of this book is to attempt to make a contribution towards rectifying this oversight through an examination of co-operative labour and its relationship with beer drinking in the Transkei. In so doing I hope also to reconcile apparently conflicting perspectives on the nature of co-operative labour in that area (Kuckertz, 1985; Heron and Cloete, 1991). Those who have discussed the festive element of co-operative work at all (e.g. Gulliver 1971; Kuckertz 1997; Geschiere 1995) indicate that it needs to be considered in relation to labour if the nature of work parties, and the relation between production and consumption, are to be properly understood. In showing how production through work parties and consumption by beer drinking are linked in the Transkei, I hope to illustrate how beer drinking and labour in Shixini form part of a single system, and thus to indicate a new approach to studying co-operative work. I do so in the context of the kinds of issues which have usually been seen as important in understanding work parties, particularly questions relating to a typology of work parties, their effectiveness and utility, the nature of reciprocity, and their historical development in specific places.

---

3 E.g. Doke (1931), Richards (1939), Krige and Krige (1943), Mayer (1951), Donham (1985), Goodfellow (1939), Gregson (1969), Kuckertz (1985), Painter (1986).

4 Some have dismissed them as 'accidental features' irrelevant to understanding the economics of co-operative work (Uchendu 1970). Some have drawn attention to the replacement of beer with other foodstuffs due to missionary influence and conversion to Christianity (e.g. Worby 1995). Even here, the socializing which concluded the work could be 'strongly reminiscent of the beer parties' (Charsley 1976).

This in turn may have important implications for one of anthropology's major concerns over the past decade, namely the study of commodities and consumption. The material on work parties and the festivities that accompany them indicates that we need to bring questions relating to production into full focus if a proper understanding of consumption is to be achieved, at least in societies where processes of consumption are not conceptually and organizationally divorced from production. Surveying the evidence on work parties presented by early Southern African ethnographers, Goodfellow commented that they were 'the principal means whereby numbers of people outside the boundaries of individual homesteads co-operate for the performing of difficult work ... the principal medium in which capital is invested for agricultural purposes, and .... one of the main groupings through which consumption is carried out' (Goodfellow 1939, 242).[5] Precisely how production through work parties and consumption by beer drinking are linked in the Transkei needs explication.

Exchange labour and festive labour

Work-parties in Africa are organized for a wide variety of purposes, in which agricultural work features prominently. Combinations of labour from a number of households arise in various circumstances in which the community is obligated to work, such as in performing tribute labour for political leaders, usually in return for food and drink, or in performing tasks on behalf of a specific community such as the building or repair of school rooms or churches (e.g. Charsley 1976).[6] Most commonly, however, labour combinations are formed on a voluntary basis to assist an individual household with a specific task, and it is primarily this kind of co-operative work that I am concerned with here.

5 The 'capital' in question is the maize stored from the previous season.

6 See also Krige and Krige (1943), Marwick (1966), Ashton (1952), Davison (1993). In southern and central Africa generally, the most prominent exchange of food for labour occurred at the chief's court, where people could come to offer their services in return for food and beer. There is a close association between chiefly authority and the provision of food, much of which is received as tribute from the chief's subjects (Richards 1939).

Co-operative labour exists in a large variety of forms, which vary according to factors such as ecology and type of economy, as well as in terms of historical and socio-cultural factors. However, two basic types of work groups have been identified.[7] The first is 'exchange labour' or the 'reciprocal work group', a relatively small group of households that work for each other regularly and in rotation on a reciprocal basis. These are often relatively corporate groups — membership is fixed and bounded, though conflicts do occur and there may be changes in the composition of the group from time to time. There may be formal leadership roles, though this kind of co-operative work is associated with relative equality of farm size and egalitarianism (Moore 1975). Typically, work of this kind is not associated with a beer-drink or other kind of festivity, though the organizer may provide everyday food and drink for the workers.

The second and larger type has been referred to as the 'festive' work group — food and drink are prepared and others are asked or invited to come and work on an appointed day to perform a specific task. The request or invitation may be made to specific people by the host homestead, or a general call for workers may be publicly made. The work is typically accompanied or concluded by the festive distribution and consumption of beer or other kind of alcoholic drink. Reciprocity is not as strictly defined as in the former type, and may be weak. These are *ad hoc* groups, disbanding after the task has been completed. The distinction between these two types is often blurred in practice and many work groups in Africa have elements of both 'festive' and 'reciprocal' types.[8] Moore contends that, formally, 'each implies and supports a very different pattern of socio-economic relationships' (*ibid.,* 273); festive work groups are associated with economic differentiation and the transfer of labour power from poor to wealthy while exchange labour is associated with egalitarianism. However, as we shall see, both types may be found in the same community (as

---

[7] This typology originates with Erasmus (1956), and was developed further by Moore (1975). See also Donham (1985) and Swindell (1985).

[8] For example, the *humwe* work parties in Zimbabwe (Worby 1995), the *Maka* groups described by Geschiere (1995), the *risaga* work group among the Gusii (Mayer 1951), as well as some of those found in Transkei (see below). Further examples are provided in Moore (1975, 273).

in the Transkei) and in such cases should be accounted for within a common analytical framework. One important implication of this is that the beer drinking which is associated with 'festive' labour is also relevant for an understanding of 'exchange ' labour.

In some areas, such as parts of Malawi and Zambia, extra-domestic forms of agricultural co-operation are historically rare (Davison 1993; Englund 1999) or are organized only 'sporadically' (Turner 1957). In other cases, for example among the Wolof, some co-operative labour was an aspect of village-wide associations, organized partly on an age and gender basis. In the Wolof case these associations were called *kompin* (after the French *compagnie*), had established leaders, and cut across kinship and class lines. They had a variety of other functions aside from co-operative agricultural work, much like age-set organizations in parts of Africa, which are often charged with communal tasks (Swindell 1985). They played a role in courtship and marriage rituals and were responsible for certain public works. They were relatively enduring groups that tried to accumulate resources with which to have festivities, and to entertain other villages (Ames 1958). Like age-sets, they are difficult to classify in co-operative work terms (Moore 1975)

Similarly, the *silika* work groups described by Charsley (1976) were clearly an aspect of religious organisation in the Marogoli area in Uganda, with each Christian congregation being associated with one or more work groups, whose labour secured funds for the church and prestige for church leaders. And Donham (1985, 145-8) describes one of the forms of co-operative work among the Maale, the *Mol?o* work groups, as being associated with enduring political groupings and as providing the context in which local politics was acted out.[9] *Mol?o* were corporate groups, enduring and relatively large, with defined leadership roles and established in terms of political interests. Working together in each other's fields gave

---

9 Both Wolof and Maale had a variety of other types of more *ad hoc* work groups that were involved primarily in agricultural work.

expression to and helped solidify those interests. Membership was restricted.[10]

In principle I am not concerned with these sorts of work groups in this book, since they appear to be derivative. For example, in the *silika* case participation was an aspect of the common activities of church members, and an important way of generating funds for the congregation. Normally the work was seen as being for the congregation, and a fee paid to the church. The fact that better farmers and members of the elite found ways of avoiding participation due to the high cost involved indicates that household production was not the main motive for organizing them. 'Working for one another was a means whereby the institution's manifest object of benefiting the church was attained, rather than an explicit benefit in itself' (Charsley 1976, 39). Other values outweighed economic disadvantages rather than, as is more usual with work groups, the economic advantage of co-operative labour outweighing the costs (e.g. of beer production). However, the dividing line between these sorts of work arrangements derived from religious or political interests, and the usual ones based primarily on the need to supplement household labour and organized on the basis of spatial or kinship principles, is sometimes difficult to draw.

Efficiency and uses

It is evident from the literature that although the homestead (which varies considerably in size and composition)[11] is the basic unit of production in most African societies, relying primarily on its own labour, larger labour combinations are necessary for certain heavy or time-consuming tasks to be successfully performed, or to be performed at all, though the nature of these tasks varies geographically and historically. This is the case even where people prefer to work individually and dislike holding work parties,

---

10 A household had to have male labour to qualify for membership (so some female headed households were left out). Newcomers, or people who had quarreled with their neighbours, could not belong. Other categories of people also did not qualify (e.g. those who did not cultivate a lot). See Donham (1985, 138ff).

11 By 'homestead' I mean a common residential unit inhabited by a group of close kin, whether a nuclear family or some form of extended family unit (sometimes referred to as a 'compound').

such as among the Ndembu (Turner 1957), where they were usually marked by complaints about the amount or quality of work done. Without work parties certain tasks would not have been done, or not done completely enough. For example, Ashton (1952, 127) reported that Basotho men did relatively little weeding in the normal course of life and that it was necessary to hold a work party to get weeding completed. The beer provided the incentive for men to participate, even though this entailed considerable expense.

Opinions differ on the economic efficiency of work parties, however, and range from Goodfellow's conclusion that for a woman to till her fields by herself instead of brewing beer for a work party would 'effect a net saving of labour' (1939, 243), to Donham's calculation of the amount of labour devoted to the beer brewing as being about half of the labour gain from the work party. Mayer (1951) and Barth (1967) also provide calculations comparing the cost of brewing (or buying) beer with the cost of hiring labour, showing that providing beer for a work party is 'economically efficient'. Making such calculations is, however, extremely difficult, especially if the value of the reciprocal labour is to be taken into account (Saul 1983), since reciprocity may be long- rather than short-term. The important point, however, is that wage labour is often simply not an option when a group of workers is suddenly and urgently needed, due to unavailability or other factors (Mayer 1951, Saul 1983). In the Gusii case, for example, workers might have been needed, say, for weeding, at the very time when everybody was occupied with this task. A day's wage could not hope to attract labour at such a time, and the householder had to rely on the moral obligation of members of his/her *risaga* work group to come and help. So the *risaga* had 'the invaluable advantage of providing a concentration of effort at the particular moment when effort is urgently needed' (Mayer 1951, 15).

But if everybody is busy with the same task how is it possible to have a work party? Although work parties may be in demand at certain crucial points in the agricultural cycle which affect everybody (ploughing, sowing, weeding, harvesting, etc.) it is often the case that fields are not all ready for a particular task at precisely the same time. The demand for labour may be at its peak for a short period, but small variations in individual fields as well as in the availability of household labour allow for the collective labour inputs to be staggered (Worby 1995). And there are considerable benefits accruing from this if it enables each field or garden to be planted, hoed or harvested at the optimum time (see also Moore 1975, 278). Apart from breaking labour bottlenecks, co-operative work groups have another advantage in that they are not specialized. They can be called upon to perform a wide variety of tasks, and reciprocity tends to be reckoned in terms of labour time spent rather than in terms of the task performed (Uchendu 1970). Clearly, work parties are much more significant than merely helping to 'speed up' the work, as has been suggested (Kuckertz 1985); for the individual household they can make the difference between a harvest and no harvest.

The sociability, rhythmic music, consumption of food and drink, supervision of work and informal competition associated with work parties have been cited by some as possibly making such labour relatively efficient as well as enjoyable (Ames 1958; Gregson 1969; Geschiere 1995), while others, probably the majority, claim that the consumption of alcohol and the lack of control over the quality of the work performed makes these labour arrangements inefficient. Saul (1983) has summarized some of these findings. Generally, the larger the quantity of beer brewed, the larger the work-party, though whether this means that the amount (and quality) of work done is directly proportional to the amount of beer brewed is an open question. In addition, the efficiency of the work is one thing, the utility of a work party quite another.

Although there may be certain economies of scale associated with work parties, the futility of attempting to make strictly 'economic' calculations about their activities seems to be recognized. How does one measure the value of turning 'days of back-breaking tedium into a brief and merry entertainment' (Ashton 1952, 131), or assess non-economic benefits such as prestige or the enhancement of social solidarity associated with festive work (Moore 1975)? Goodfellow understood this many years ago, writing that 'the economics of the work-party ... are simply not directly utilitarian, and show, incidentally, that production is no more 'direct' among these very backward people than it is in Western Europe and North America ... Thus another text-book generalization as to the differences between 'primitive' and 'advanced' economics must go by the way of all such pronouncements' (1939, 243).

The necessity for co-operative labour in rural areas also arises in situations where labour is scarce (e.g. where people are absent from their homes as migrant workers), where poverty creates a shortage of other resources, such as draught animals or agricultural implements, and where either cash or hired labour are in short supply (for example, where all residents have land and are cultivators) (Moore 1975; Weinrich 1975; Swindell 1985). But the form that such labour takes is governed not only by these kinds of factors, but also by social, historical, ecological and cultural considerations, topics we turn to below.

The question of how efficient work parties are is also relevant to development issues. Mayer (1951) felt that Gusii work groups would lend themselves to being used by the administration for implementing agricultural reforms, and indicated that progressive farmers growing cash crops still made use of reciprocal labour. Painter (1986) considers the possibility of making use of West African work groups in development projects initiated from outside of the societies, but his findings are inconclusive. It is likely that earlier forms of labour co-operation are in fact not suited to the constraints of more modern, co-operative production instigated by an external agency (James 1985), whether of a capitalist nature or in terms of an ideology such as *ujamaa* (Swindell 1985). The complex dynamics of changing forms of labour in relation to other kinds of changes such as the introduction of cash crops indicate that this issue has to be viewed cautiously (Worby 1995; Geschiere 1995; Englund 1999). Survey results in East and West African countries show that there is little reliance on indigenous forms of co-operative work among 'progressive' farmers (Uchendu 1970) and thus support the view that work-parties are not often particularly compatible with 'development'. However, this is likely to vary according to what is meant by the latter.

Who works for whom, and why?

In most of the published ethnography, the individual homestead or household (terms used interchangeably here) is regarded as the basic unit of agricultural production, using the labour available within the household. There are exceptions, however. In the case of Darfur, every individual worked for him- or her-self (Barth 1967), relying heavily on beer for labour exchanges in the productive process. Among the matrilineal Ndembu, too, agricultural production was predominantly individualistic. Men and women each worked on their own and did not have rights in each other's gardens. Work parties, though infrequent, were performed for individuals and not for households (Turner 1957, 23). In other cases a woman might have a private field or garden in addition to being part of a productive unit based on the household (e.g. Saul 1983; Weinrich 1975). Generally, voluntary work groups in Africa are organised by an individual homestead for the benefit of its members as a unit. The questions which arise here, and which revolve around the matter of interest, are: What is the basis of the relationship on which homesteads provide labourers for others? Which homesteads organize work parties? What motivates both hosts and workers? I believe that a proper answer to these questions in the Transkei case requires a close examination of the various forms of co-operative labour in relation to each other, and of the festivities with which some forms of this labour are concluded.

Since productive activity is carried on by households for their own individual benefit, and extra-domestic labour is provided by other households, we have to concern ourselves with the relations between households rather than with relations of production within the individual household, which are generally based on factors such as age and gender and shaped by social norms, rights and obligations (the social division of labour). But this does not mean that we can ignore the internal relations, since they have a crucial bearing on the need for co-operation between households in the first place and on the relationships through which this labour is organized. This will become evident particularly in the historical circumstances that have a bearing on the development of co-operative labour arrangements, both in the Transkei and elsewhere in Africa. We have therefore to do two apparently contradictory things — we have to treat the household as the basic unit of production, and to understand co-operative labour as co-operation between members of different households, *and* we have to understand this in terms of the internal structure of individual households and the historical developments affecting that structure. This is addressed more fully in relation to Shixini in the next chapter.

In some cases it is evident that labour for work groups is recruited primarily or exclusively on the basis of kinship, as in the case of *Ganyu* labour described by Englund (1999), but this is exceptional. The evidence shows that spatial rather than kinship relationships are usually crucial in determining the pool from which the households providing labour for the organizer are drawn. For obvious reasons it is primarily neighbours, people within the same village or hamlet (or other territorial unit), who normally provide the labour required. Of course, since people who live in the same vicinity often tend to be kin as well, ties of kinship and neighbourhood reinforce each other and both contribute to the solidarity of work groups. This was the case with the Gusii *risaga* work groups, a form of relatively permanent labour organisation formally based on geographical or neighbourhood divisions (Mayer 1951). Where lineage segment members occupied a common territory, the members tended to form a single *risaga* group and to exclude non-kin. Some groups were known by the name of the kinship group involved, though most were named after the geographical area that they occupied. Mayer felt that kinship was not necessary to the formation and maintenance of these groups, which were based primarily on 'neighbourhood and economic independence' (1951, 10).

Painter (1986) describes a similar scenario in southwestern Nigeria, though here, too, one form of co-operative labour that had existed in the past was based on kinship. In the Maale case, territory provided the formal basis of the pool from which co-operative labour would be drawn, with kinship and affinity being important in defining the nature of this pool, though this had also to be seen in a historical and political context. Within the pool, however, kinship was no more likely to be used as a basis for labour-cooperation than neighbourhood and friendship (Donham 1985). In some cases co-operative work takes place within a kinship idiom and the significance of kinship relations seem to be directly related to the fact that people live near each other and work together (Gulliver 1971). In many parts of Africa, however, participation in a work party is not based on any ascribed relationship, whether based on kinship, territory or political status, but is voluntary (e.g. Saul 1983; Worby 1995), though practical considerations mean that it is usually neighbours or people from the same village or other territorial unit who collaborate.

Bloch (1973) has argued that kinship relations are characterized by strong morality and can tolerate long-term imbalance in exchanges. Other relationships, based on politics, friendship or neighbourhood, 'have shorter term than those classed as kinship and thus are less moral' *(ibid.,* 77) though he is 'quite prepared to believe that in some societies other types of relationships seem to have equal if not 'more' moral commitment than kinship' *(ibid.,* 77). Since kinship relations have a high moral content the effect of involving kin in co-operative work is to make for long-term tolerance of imbalance. Relationships based on factors such as territory or neighbourhood are less moral, and tend to lend themselves to short-term interest and more immediate reciprocity. Kinship links are more reliable and long term and need not be constantly acted upon. Reciprocal relations are maintained with non-kin (rather than kin, between whom imbalances are tolerable) in the form of co-operative work arrangements so that a large pool of potential collaborators is maintained. Bloch interprets the Merina tendency to create artificial kinship links with the neighbours on whom they depended for agricultural co-operation, and also Gulliver's (1971) work, which indicates that the significant Ndenduli kinship relations are based on the fact of working together, in this way. The implications of his argument are that the economic 'interest' in labour exchanges is likely to be more prominent where the relationships are less moral and more immediately reciprocal. The implication is not, however, that more moral relations (based on kinship or otherwise) are more interest free, simply that they can tolerate longer-term imbalance.

Whether based on kinship or not, the fact that forms of co-operative labour disappear or become modified is contexts of resettlement, migration and other forms of social upheaval where people find themselves living with relative strangers (e.g. Wilson et al. 1952; James 1985; Worby 1995), indicates that work parties are associated with a socially well integrated community in which there is mutual trust and people know each other well enough to ensure that reciprocal duties are exercised.

In Southern Africa the evidence indicates the primacy of neighbours in co-operative labour (Shaw 1974), though territorial units are seldom *formally* used to define the boundaries of work groups, and precisely which neighbours are asked or invited to work is less clear but of obvious importance (Kuckertz 1985). Marwick (1966, 166) says that a Swazi man organizing a work party sends word to his 'more industrious neighbours' asking them to come and help. In Lesotho, people invite their 'close friends and neighbours' Ashton (1952, 131). However, the Transkei evidence indicates, as we shall see, that certain kinds of work groups tend to be composed primarily of close kin from within the neighbourhood, while in others kinship seems to be of little significance, and the implications of this need to be examined.

Work parties and rural differentiation

Many commentators feel that 'festive' labour tends to involve transfers of labour from poorer to wealthier households, or to political leaders, and that 'exchange' labour tends to be found among households of relatively equal economic status (e.g. Swindell, 1985). However, the evidence on what kind of households benefit most from African work parties, and whether these are of a certain kind or not, is not conclusive, and probably can never be, given the many different forms of co-operative labour and the variable conditions under which work parties arise and operate in different parts of the continent. Imbalances of various kinds have been reported — sometimes one territorial group, one class of households, or one type of kin providing labour more frequently than another (Gregson 1969; Painter 1986).

In a number of cases it has been argued that where rural differentiation is marked, work parties represent a net transfer of labour from poorer to richer homesteads, in that it is the wealthier homesteads which can afford to brew beer and provide the food that is required to host a work group (e.g. Mayer 1951; Donham 1985, 163). However, the conclusion that this allows wealthy homesteads to remain wealthy (because they can ensure that they get labour when they need it) and locks poorer households into poverty because they are unable to cultivate at a favourable time, or at all, as a result of their labour service to the wealthy, is not necessarily generally applicable.[12] This position rests on a variety of assumptions — that reciprocity is not strict and need not take the form of labour, that the reward provided to the workers is a payment in kind, and so on, which have to be examined in each particular case. There is evidence to suggest that such assumptions are not always warranted, in Southern African communities as well as elsewhere in Africa. This issue is discussed more fully below in the section on reciprocity.

It has also been claimed that the norms governing co-operative work can prevent wealthy households from increasing their wealth at the expense of their poorer neighbours (Saul 1983; Geschiere 1995), that co-operative work is a means of redistributing wealth from wealthier to poorer households (Englund 1999), and that without co-operative labour institutions, poorer households would simply not be able to survive or to manage certain vital tasks (Shaw 1974; Sansom 1974). For example, Richards pointed out that a destitute Bemba woman could offer labour as part of a work group and receive much needed food in return. Women who were deserted by a husband, or whose husband was absent from home, could get heavy work done (such as building a granary) through attending work parties, using the maize obtained as reward to brew for her own work party to get the task done. This was the only way, in fact, that such work could be done under the circumstances (1939, 144-6). The ability of poorer households to exchange their right to a work party for cash, the workers performing their duty on the buyer's behalf (Uchendu

---

[12] James (1985), for example, drawing on Webster's material on the Tembe-Tonga as well as her own data from the Northern Transvaal, makes such an argument.

1970; Geschiere 1995), is possibly a further mechanism by which poorer households benefit from a reciprocal labour system, though it could also be argued that this is a way of enabling the wealthy to secure surplus value.

Sansom (1974) has suggested that reciprocal work parties allowed all households in a community to have more or less equal access to available labour, and that this led to a degree of egalitarianism which was manifested in grain yields. In the case of the Pedi of South Africa this was related to the centralized authority of the chief, whose permission had to be obtained before ploughing and planting (and thus work parties) could commence. The chief could also decree an end to such activities, ensuring that some households did not cultivate more than others. However, the overall evidence suggests that we need to be careful of associating co-operative labour with an egalitarian ethos (Worby 1995), since it is often co-incident with differentials in wealth and unequal access to resources. Sansom's argument, if correct, applies only to the pre-colonial situation when it was difficult to accumulate more grain than others because land was not scarce, and work party labour was based on reciprocity (*ibid.*, 154). In addition, it clearly applies only to commoners, with chiefs probably keeping a watchful eye to ensure that no commoner became too wealthy.

Reciprocity

The issue of who participates and of the motives for participating in work parties is closely tied to the question of reciprocity. All over Africa, work parties are described as being governed by a norm of reciprocity,[13] as implied in the term 'reciprocal' work groups. In the case of 'festive' work parties, however, the nature of reciprocity is not always clearly defined by the participants. In some cases it may be that the reward offered at the end of the work, whether it be in the form of food or drink, ends the transaction (e.g. Barth 1967), but more frequently the host homestead is also expected to reciprocate with labour at some point in the future. The beer

13 E.g. Shaw (1974), Hunter (1936), Wilson et al. (1952), Mayer (1951), Uchendu (1970), Donham (1985), Geschiere (1995), Worby (1995).

and meat are valuable, there is a 'definite exchange of work and food involved' (Richards 1939, 146) and the amount of work performed is related to the amount of beer/food provided, but 'reciprocal work is also essential' (Goodfellow 1939, 246). Food and beer are not sufficient reward; labour has to be reciprocated with labour, but the expectation of labour reciprocity by itself is not sufficient to induce people to work.[14]

The Gusii *risaga* group described by Mayer had decidedly 'festive' elements although it was of the 'reciprocal' type; it was an enduring group with fixed membership, and any member of the group could call for a work party when it was needed. The member who did so was obliged to brew a considerable quantity of beer each time, but this did not end the transaction since 'behind this immediate reciprocity of beer for work lies a wider reciprocity of help for help. To give ones labour, at the price of a beer drink, is a mutual obligation' (1951, 6). Karp has made a similar point about Iteso work parties. The beer that is provided for workers is not an exchange of beer for labour, but facilitates and is part of 'a complex series of [delayed] reciprocal exchanges ... the vehicle through which co-operation is achieved' (1980, 88). Precisely how beer drinking articulates with labour reciprocity, however, has not been the subject of analysis beyond rather vague assertions that it indicates sociability and the willingness of people to work (e.g. Karp 1980). As we shall see below in discussing the Transkei case, this articulation is directly related to an understanding of the reciprocity associated with work-parties and to the development of a unified analytical framework within which to make sense of a variety of types of co-operative work

Even in 'festive' work arrangements, labour reciprocity can be strict. In Zimbabwe, participation in *humwe* work parties is governed by the rule that 'for every participant sent from one household to work at the *humwe* of another, an equivalent number must be sent in reciprocation ... Equivalence is reckoned in units of adult labour' (Worby 1995, 14). Participation is supported by an ideology of mutual helpfulness. But it is partly because of the strictly reciprocal nature of these groups that wealthy

[14] See, for example, Sansom (1974), Gregson (1969), Wilson et al. (1952), Saul (1983).

households stand to benefit from them more than poorer households in the long run. A large, well-developed household can send a relatively large number of people to join the *humwe* groups organised by others, creating a large number of reciprocal obligations. Its productive capacity is enhanced through the calling in of these debts, when it can raise a *humwe* consisting of a large number of people. Smaller or newer households that cannot supply much labour for others can also not build up many labour credits. *Dabo* work groups in Maale operated in a similar fashion (Donham 1985). However, elsewhere in Zimbabwe, and in other parts of Africa, it is the expectation of reciprocity (the opportunity cost of their own labour) as well as the need for more efficient and accurate work that makes wealthier or commercial farmers avoid work parties (Moore 1975; Weinrich 1975, 88, 264; Charsley 1976).

The sanction for non-participation confirms the principle of reciprocity, in both types of co-operative labour. Generally, those who cannot attend a work party which they might be expected to join should give a reason why they cannot be there, otherwise they are likely to be refused assistance when they themselves need help, or to receive it only reluctantly (Gulliver 1971). People who persistently shirk their responsibilities can in extreme cases be ostracized and forced to settle elsewhere.[15] Imbalances may be tolerated, especially between kin, and justified in terms of the moral obligation to help kin at a critical stage of their household development (Donham 1985, 164). Or reciprocity may take forms other than labour, as when wealthier or more powerful households provide political support or other favours for those who labour for them (*ibid.*) Many work groups operate under a general assumption of reciprocity and mutual helpfulness, in that people help each other without expectation of immediate return, but in the knowledge that the receiver could be asked to provide labour, or grain, or something else, in the future, perhaps in the next season. It is a form of social insurance (Sansom 1974). However, this general expectation operates within a pool from which labour is drawn,

---

15 See, for example, Kuper (1947, 147); Mayer (1951); Wilson et al (1952); Donham (1985).

which must be distinguished from the actual relationships between people in terms of which they give and receive help.

The notion that work parties are governed by an ethic of reciprocity and mutual helpfulness has been disputed by Kuckertz (1985), in a detailed analysis of one form of co-operative work in Pondoland, approximately 200 kilometres northeast of Shixini. Historically an independent chiefdom, Pondoland is inhabited by Xhosa-speaking people culturally very similar to those in other parts of the Transkei. It is necessary to examine this claim, because, if Kuckertz is correct, the argument presented in this paper may be misconceived and the moral basis of co-operative work in the Transkei might lie not in notions of mutual help and reciprocity but elsewhere. In addition, an analysis of Kuckertz's argument may provide further evidence of the need to locate work-parties and their associated morality within a wider, changing socio-economic context.

Kuckertz looks at the *ilima* work group, consisting of around 20-25 people, in terms of which an individual household requests help from a number of others, especially in the hoeing season. It is typical of Southern African work parties, elsewhere called *matsema* among Basotho, *lilima* among Swazi, *lejema* among Lovedu, etc. In Caguba village of the Mthwa chiefdom where Kuckertz worked, there is no public announcement of *ilima*; the homestead head (or his/her delegate) simply goes around asking (*ukucela*) people to come and work. Participation is in the interest only of the organizing homestead, which brews beer for the workers. People participate out of the moral obligation to assist those who are in need of help and who ask for it. Work parties are not based on any notion of generalized reciprocity because Mpondo themselves do not see it in these terms. People participate not because of economic factors but because of moral ones, and joining a work party is not to the volunteers' economic advantage since the recipient is not obliged to reciprocate with an equivalent labour input or in any other way.

The head of the organizing homestead cannot rely on a general assumption of mutual helpfulness; he cannot assume that people will know that he is having a work party, and arrive to assist him in the hope of receiving assistance in return. He has to ask for help. The fact that there has to be an explicit invitation, and that the invitation is selective, casts doubt on 'the assumption of a generally operating system of mutual helpfulness and reciprocity, and highlights the fact of individual interests...' (*ibid.,* 116). People come along out of 'long-standing friendship' (*ibid.,* 122), although some attend on their own initiative without being asked, having heard of the work party from the invitees. Work parties take place within social contexts where relationships exist independently of purely economic calculations. Who comes to help has little to do with neighbourhood proximity, kinship or membership of the 'mat' associations (hospitality groups). Friendship may be based on any number of factors, including kinship, common initiation, or previous assistance rendered. So the people who work are part of the homestead's network of relationships. It is an action set. However, there are territorial limits involved. In the example Kuckertz provides, all but three of the 26 people who came to help were from the same ward or village, the other three from a neighbouring village, and 15 were of the same mat association.

Work parties thus 'underscore the economic individualism typical of Mthwa ... rather than being instances of co-operation in the sense of sharing economic interests. *Common* economic interest ... exists only on the level of the individual homestead' (*ibid.,* 117, emphasis in the original). People would prefer not to have work-parties because of the expense of providing beer 'which cuts deeply into the resources of the organizing household' (*ibid.,* 120) and because there is a risk of trouble and tension (over issues such as adultery) at the beer party. Yet Kuckertz records that in the 1980-81 season 85 out of 106 households organised work parties, mostly related to the cultivation cycle.

Kuckertz is correct to characterize production in Pondoland as one of 'homestead economic individualism' and to regard work parties as being in the economic interest of the sponsoring homestead. It is like that in Shixini too. But he seems also to assume that the two kinds of interest (individual and common) cannot co-exist and are mutually exclusive (though at one point he does refer, without elaborating, to 'group interests' (*ibid.,* 115)). In the Transkei, common economic resources are represented in things like grazing, water, thatching grass, building materials, and so on. These things are held in common by territorial groups; they are of obvious economic importance and utilized by individual households. There are rules governing their use, indicating that individual usage should not be at the expense of others in the community. Secondly, if homesteads depend heavily on inter-homestead co-operation at certain times of the year, as Kuckertz rightly claims, then it is in the economic interests of each homestead for other homesteads to continue to exist and for every homestead to maintain friendly relations with a number of others. If each homestead depends on the support of others it is in the common interest for all homesteads to survive and grow. Only by trying to keep the social and the strictly 'economic' separate can this be ignored.

If there is no norm of reciprocity and assumption of mutual helpfulness, asks Kuckertz, how is social cohesion maintained, how does the society hang together? It hangs together because the organizer of the work party will at a different point in time help others when he/she is asked; he/she, too, adopts the moral principle of helping those in need who ask for help, and is obligated to accept this principle since he/she acted in terms of it in asking for the help of others (*ibid.,* 128-9). Long and short-term dimensions (morality and interest) of the issue are interwoven to provide social unity over time. But it is morality that is the motive, not reciprocity. Yet it is hard to reconcile this with the conclusion that 'the heavy engagement in work-parties during a certain period of the year creates an infinite number of long-term relationships. The short term economic relationship comes to an end with the beer drinking after the work party has fulfilled its task. The workers can go home freed from the burden of the promise they gave. The homestead head ... is left with the lasting obligation inherent in the moral precept which gained him his economic advantage' (*ibid.,* 129).

Caught in an unsuccessful attempt to keep social and economic domains separate (*ibid.,* 115), the result of Kuckertz's analysis, although aiming to examine the effects of motive, is to confuse motive with effect, by specifying the moral (emic) conditions which underlie the analytical notion of reciprocity. Reciprocity in this case is the analyst's tool, it need not be stated explicitly by the people involved. The shared ethic of giving help that is rightfully asked for has the long-term effect of producing an ethic that implies reciprocity and mutual helpfulness. So, far from demonstrating that reciprocity does not exist, he specifies the moral conditions that underlie it and make it necessary. It seems that part of the difficulty here is that a distinction is not made by Kuckertz between a general and abstract moral precept such as 'mutual helpfulness' which admittedly has no specific explanatory power, and reciprocity as norm which, because it presupposes action (resulting from or in anticipation of another action), is compatible with specific interests and can be used to explain behaviour in a specific interactional situation (Iturra 1977). Gulliver (1971, 214)) seems to be referring to this in speaking of 'the practical morality of reciprocity' in his analysis of cultivating groups among the Ndenduli in terms of interest, networks and action sets.

Why should Cagubans not express any ethic of mutual helpfulness and reciprocity, prefer to do without work parties, be reluctant to brew beer, and therefore be so different from the people described by Hunter in Mpondoland in the 1930s (and also from Shixini people today)? The question is left unattended by Kuckertz but it needs to be addressed in the context of the overall theme of this paper. It involves a look at some of the major social and economic changes that have affected the lives of Cagubans since the time of Hunter's fieldwork and which may have had the effect of changing moral norms and fostering the development of an individualistic ethic, with consequences for how people construct their reliance on co-operative work.

One of these is missionary influence, conversion to Christianity and increased education and westernisation (which are often associated with conversion). Kuckertz points out in a footnote that 'Mpondoland has been under the influence of Christian preaching for about 150 years' (*ibid.*, 130). However, only around 4-5% of Mpondo were Christians when Hunter worked there in the 1930s (Hunter 1936, 349), whereas 88% of Cagubans were Christians when Kuckertz did fieldwork (Kuckertz 1983/84, 114). Evidence from other parts of Africa indicates that conversion to Christianity coincides with changes in the nature of co-operative work (e.g. Worby 1999; Charsley 1976; Mayer 1951; Wilson et al. 1952) partly because of the prohibition on beer drinking that goes with conversion. And in Pondoland itself Hunter pointed out that although *amalima* work parties were 'very extensively used' in more conservative districts (1936, 90), with virtually every homestead having at least one a year, mainly for the first weeding, Christians or 'dressed' people held *amalima* infrequently, partly to avoid beer-drinking, partly because they were 'more industrious' and attempted to maximise production, i.e. they were more driven by utilitarian interests (*ibid.*, 91).

A second important factor lies in the agricultural betterment schemes widely implemented in Transkei, including Caguba, in the 1960s and 1970s. It is well known that these schemes, usually involving the relocation of people into new, concentrated village-type settlements, where they have had to establish new homesteads with minimal or no compensation from the state, have had dire economic and social effects, including the further entrenchment of migratory wage labour, the breaking down of generations old neighbourhood groups, and the creation of suspicion and ill-feeling between new neighbours. They severely disrupt the social order, and in many cases it has been reported that patterns of economic co-operation have fallen away with the introduction of these schemes (de Wet 1989; Mayer 1980; McAllister 1989). Moore (1975, 282-284) has specified the general conditions under which one would expect co-operative labour to decline, and in the case of betterment, all of these conditions were met. James (1985), for example, has illustrated how the implementation of a betterment scheme in the northern Transvaal, together with the introduction of a state-run co-operative, helped to transform agricultural labour. Before the changes, labour was based on reciprocity, but this eventually gave way to contractual relationships. In addition, relocation in terms of the betterment scheme allowed newcomers into the area. Many of these were converts to Christianity and disdained work parties because of their association with beer. Others, not originally part of the community, lacked the social ties to successfully join the reciprocal work party exchange circuit that continued, at first, to exist.

Betterment villages usually provided for very small homestead and garden sites, and attempted to ensure that arable land was given to all homesteads, thereby reducing the size of land holdings. It is likely that it is because of betterment that Cagubans have such small fields (average size of only half a hectare), when compared with the pre-betterment situation, when fields were much larger in Mpondoland (Hunter 1936, 72-73). This, together with the increasing reliance on migrant labour that betterment invariably brought, may have had important effects on the role of agriculture in the local economy and on the nature and importance of co-operative work. In short, it is likely that homestead production declined and became relatively unimportant for economic survival, with migrancy and wage labour growing in importance. Kuckertz seems not to recognise this and says that migrancy 'has no direct implications for the organization of work parties' (1985, 118).

Shixini people, by contrast, have been relatively unaffected by changes of this kind. They too depend heavily on migrant labour, but until the early 1980s many men were able to spend long periods at home between spells of wage employment, and even today over 80% of migrants visit their homes at least once a year, often at the start of the agricultural season. There has been little Christian influence in Shixini, the area is relatively isolated, and betterment, as we have seen, has not been implemented.

5

Work parties and the transformation of rural production

A number of writers have pointed out that co-operative work groups cannot be seen simply as indigenous or traditional responses to the need for collaborative effort.[1] To do so 'is to infiltrate a perspective. The institution so described comes readily to be seen as an entity 'persisting' from an unchanging past, to be sharply distinguished from the new and the modern ... it's persistence or decline taken as an indicator of development change, but any sense of it as a dynamic aspect of society and economy is quickly lost' (Charsley 1976, 34). Instead, work parties are responses to particular situations, and have to be placed within a historical, economic and cultural context. In this section I consider some of the recent analyses of co-operative work, which show the necessity of this kind of approach, before applying some of these lessons to the Transkei case.

Charsley's warning about unilineal perspectives seems now to be well established. For example, it has been stated that some cases co-operative work of the large, 'festive' type may have arisen with the decline of slavery in Africa in the late nineteenth century (Swindell 1985, 138ff) espe-

[1] E.g. Painter (1976), Charsley (1976), Donham (1985), Worby (1995), Geschiere (1995).

cially as this occurred more or less conterminously with the development of commercial export crops and the need for additional labour sources. Its development may therefore not only be quite recent, but one cannot make assumptions about its lineal development from, or into, other forms of work, including wage labour. To take a somewhat more developed example, Geschiere (1995) shows how among the Maka, a new form of co-operative work group developed as cash cropping of cocoa and coffee were introduced (in contrast to other cocoa producing areas, where wage labour developed). Although wage labour was known, there was strong resistance to it in Maka villages, and the form of labour that was adopted was based on reciprocity and the equality of members. This was not just the adoption of a 'traditional' form, but arose as a specific solution with specific labour demands, given the socio-cultural context of production, in which the key element was that individual ambition co-existed with an egalitarian ethic. Reciprocity in work limited the ability of the rich to become richer, but great importance was attached to the festive element, the 'reception' to reward the workers, which if lavish and successful could enhance the host's prestige. Although wealthy farmers could benefit from hosting work groups, they had to do so within a framework in which reciprocity remained the norm, and poorer farmers could rent out their turn to host the group for cash and thereby maintain their farms.

Nor should the presence of waged work necessarily be interpreted as a change from 'tradition' and a move towards contractual and individualized social relations, as a recent analysis of *Ganyu* (paid) work in Malawi demonstrates (Englund 1999). In Malawi's Dezda district, *Ganyu* labour is the most common way of recruiting agricultural labour from outside the household. Changes in labour arrangements do not necessarily mean changes in the nature of society and morality, however, and just because money is involved does not mean that these labour arrangements are not linked to 'social obligations and privileges' (1999, 138). Dezda households prefer not to work collaboratively but to rely on their own labour (see also Davison 1993). To Englund, this represents an institution of the *longue duree*. Work-parties of all kinds are rare, and participation in them is not based on reciprocity, but is rewarded with beer, which ends the transaction. The ability to provide the beer for such an occasion indicates that the household is relatively affluent, but even such households prefer *Ganyu* labour, where the workers are selected by the householder and paid. But these labour arrangements take place in accordance with the household head's constitution as a person, i.e. in terms of his/her position within the household, relations with kin and affines, and dependence on the land, which it holds through residence in a particular place and association with a matrilineage, and from which it obtains the essential maize. The person is 'made' by maize as the basic foodstuff, maize that is available by virtue of residence in a particular territory and location in a kinship group (Englund 1999, 144).

It is kin who are employed, particularly in agricultural tasks. Households with cash income from employment or which expect better harvests than others are expected to use *Ganyu* labour. Prominence is measured by the extent to which people are able to recruit others to work for them, and wealthy or prominent people are expected to benefit their kin in this manner. The workers are from individual households, recruited by an individual household, the relationship thus making visible distinct households and the nature of the recruiter's household as one constituted by a particular set of relationships and entitlements (e.g. to land). Thus '*Ganyu* labour arrangements are integral to the constitution of economic actors as moral persons' (*ibid.,* 139) and 'the interests involved in selecting *Ganyu* labour can be understood only within the wider context of relationships which constitute persons are composite selves' (*ibid.,* 151).

The recruiter's interest is to protect and increase wealth, but morally, the protection and accumulation of wealth has worth only if it is done through mobilizing others. 'Wealth which is individual and private, mobilizing no one but the person him- or herself, constitutes its proprietor as the inversion of the moral being, a witch' (*ibid.,* 151). Furthermore, the production, distribution and consumption of material goods cannot be distinguished from social reproduction, since these material things 'make a vital contribution to the very making of persons and households' (*ibid.,* 140)

Of course, things are not so straightforward in real life, which is always beset with conflicts and contradictions. One cannot assume consensus and an absence of self-interest (*ibid.,* 141). Since the self is a 'composite of relationships' one has to place the individual actor within the social network of which the actor is composed. But the problem is that a recruiter has to choose which kin to involve in this way, and that those recruited choose to do the work. The latter, too, are liable to be accused of witchcraft 'if seen to be passive recipients of others wealth' (*ibid.,* 152). If inattention to certain relationships generates conflict, this can mean that the individual's interests are re-configured, patterns of recruitment and work adjusted. The generalized moral sentiment is one thing, the way in which it is acted upon is another, and it is in this respect that the analytical notion of the composite self is useful 'for tracing the sources of normative dissonance amid a generalized moral sentiment' (*ibid.,* 152); 'The analytical problems of self interest and labour arrangements require an analytical notion of the composite self. It is a tool with which to discern the origin of interests in a multitude of acts and contingencies, none of which is reducible to any one actor' (*ibid.,* 146). Englund's analysis has important implications for an understanding of co-operative work in Transkei, despite a different historical and cultural context.

The 'commoditisation' of labour in African rural areas is frequently associated with a decline in earlier forms of collaborative work and the use of wage labour when it becomes necessary to secure workers from beyond the household. Davison (1993), for example, argues that women in southern Malawi refrain from organizing collaborative labour because they are concerned to maximize their production, both for subsistence and for cash sale. In any case, the 'cultural tradition' of collaborative production in these areas is weak. Englund provides an obviously different perspective on this in one part of southern Malawi, and Worby (1995) has warned against unilineal models of rural change which see wage labour as emerging from older forms of collaborative production as rural dwellers turn to cash cropping and become more proletarianised. In rural Zimbabwe, he argues, wage labour relations go back to the early colonial period, and should not be viewed as the antithesis of collaborative labour.

He stresses the need to analyse different forms of co-operative work in relation to each other. Wage labour in some parts of rural Zimbabwe co-exists with co-operative labour, and we need to make a careful study of how they relate to each other in particular cases.

In the Gokwe area studied by Worby, he found little evidence of communal (as opposed to kin-based) work groups until the 1950s, when agricultural production increased due to a number of factors and large labour combinations became necessary. Community based, strictly reciprocal work groups called *humwe* were organised to meet labour needs. The nature of these groups was transformed with the introduction of cotton as a cash crop in the 1960s. Cotton production required a higher quality of labour, and *humwe* groups now became 'commoditised' in that beer was replaced with items such as tea, sugar and other shop-bought goods. Later, a goat or sheep were provided as well. Worby considers the quality of earlier *humwe* labour to be poor because it was performed 'under the debilitating influence of alcohol [i.e. maize and sorghum beer]' (*ibid.,* 15). Those who first started growing cotton were churchgoers and thus did not drink beer, and so 'the beer party became a tea party' (*ibid.,* 16).

At the same time, work parties based on kinship, where reciprocity was not strict and which were thus not suited to the strict requirements of cotton production, were transformed into more strictly reciprocal arrangements whereby two or more households agreed to work for each other on a rotational basis. Both kin and non-kin were involved in these arrangements, but the terms of the agreement, based on a careful calculation of labour inputs, were not defined in kinship terms. People who tend to make use of this form of labour are frequently immigrants or people who have been resettled and probably stems from their 'need to establish relations of exchange and co-operation with strangers in order to access labour or means of production' (*ibid.,* 18). The introduction of cotton also gave rise to a dramatic increase in the amount of individualized casual wage labour, also strictly contractual. However, this did not lessen the incidence of collaborative, reciprocal labour, either of the earlier, communal *humwe* type or the type that arose out of kinship co-operation. On the contrary, such forms of labour are on the increase. Virtually the only people who employ casual wage labour are themselves salaried or self-employed as traders or vehicle-owners, not peasant farmers.

These three case studies, cited here initially as argument against unilineal views on the development of labour forms in Africa, provide evidence of the need to understand the historical conditions in which co-operative labour arises and of the need to analyse its relationship with other forms of labour in the same society. The lesson that it is in the changing context of rural production that the development or dissolution of co-operative work should be understood seems a relatively obvious one, but the varied nature of rural, historical and socio-cultural conditions in Africa make it anything but a straightforward exercise to apply.

The historical context: changing rural production

Turning to the contemporary context of co-operative work in the Transkei, it is apparent that a historical perspective is required in terms of which work parties can be seen as linked to both the relations between households and the internal composition of households in general. In this section I rely primarily on William Beinart's analysis of the changing nature of rural production in Pondoland, supplemented with evidence indicating that it applies also to other parts of coastal Transkei, which is culturally, ecologically and historically (at least since annexation) relatively homogenous.[2] Beinart's analysis is supported in many respects by Hunter (1936), and its relevance to other parts of the Transkei demonstrated by Wilson et al. (1952), Wilson (1971), Hammond-Tooke (1962), (1975). His observations are in any case not confined to the Pondoland districts. The analysis revolves around changes to homestead size and composition, the institutionalisation of migrant labour, and other related factors, which combined to produce a range of institutional adaptations, and which in some ways altered the nature of rural Xhosa-speaking society. The environmental and ecological factors relevant to homestead production in pre-colonial times have been sketched in chapter two, above.

2 A fuller summary and interpretation of this evidence is found in McAllister (1985).

Here I concentrate on the political and socio-economic factors that impacted on homestead production after annexation.

In pre-colonial times in Transkei (including Pondoland), homesteads were large and polygynous, consisting of between ten and forty huts. A homestead was made up of an extended family, under the control of a patriarch who provided the cattle from the homestead's herd to enable his sons to marry. His various wives were associated with different 'houses', each house associated with property allocated to it by the head and to which it had a right, including livestock and lands (fields and gardens). Labour was provided by members of the household — primarily the head's wives, their children, and his son's wives.

Homesteads became smaller in the period roughly between 1900 and 1940.[3] Migrancy was partly responsible for this. As early as the 1920s in Pondoland, slightly earlier in other parts of the Transkei, maintaining rural production depended partly on access to cash through migrant wages. Diseases had wiped out most of the cattle, and wage-earnings were needed to replace them. There were other cash needs also, e.g. for taxes, as rural homesteads got drawn into the developing capitalist economy. It became common for one or more of the young men in each homestead to spend some time on the mines and other industries earning a cash wage. By 1936 already, in Pondoland, the large majority of homesteads had at least one migrant, and 45% of men between 15 and 45 years of age were absentee wage earners. This rate was higher in other parts of the Transkei (Beinart 1982, 94-95).

Migrancy led to tensions between older and younger men in the homestead, largely over the control of the migrant's wages and the cattle purchased from it. Younger men were no longer dependent on their elders for marriage cattle and could arrange their own marriages instead of depending on arrangements made by their fathers. Growing individualism due to education and contact with missionary ideas also contributed to a change in the relationship between fathers and sons, and sons started to establish

---

3 This was a widespread trend among Xhosa speakers in general; see Hunter (1936, 25); Wilson et al. (1952, 52-9); Hammond-Tooke (1962, 35-6); Wilson (1971, 63); Hammond-Tooke (1975, 82-3).

their own, independent homesteads earlier in life than before. The senior generation also started to play less of a role in land allocation as population pressure grew and land became scarce and this task fell to the headman of the location. Since land was allocated to homesteads, having their own homesteads made it easier for the younger men to get fields. New technology, principally the plough and the introduction of ox-draught, enabled the smaller homestead to cope. The migrant became both wage earner and independent homestead head. Although this meant a loss of power to the seniors it was not always opposed by them, partly because it freed them from responsibility for civil liabilities incurred by their sons, since the latter were homestead heads in their own right. Other factors contributed to this process. Homesteads also became smaller due to the decline in polygyny arising from cattle shortages and the influence of missionaries, and because there was no longer any need to maintain large homesteads for defensive purposes (*ibid.,* 97).

This move towards smaller homesteads had important implications for the organisation of agricultural production. Most importantly, maize replaced sorghum as the staple crop. Sorghum is vulnerable to birds and needs to be protected as it ripens, and the smaller homestead did not have the labour to do this because children were too young or in school and the husband was often away at work. Maize develops within the protection of a sheath and does not require labour to guard it from birds. It is also more amenable to inter-cropping. In pre-colonial times, vegetables and grain were cultivated separately, with vegetables being grown in gardens near homesteads and grain in more distant fields. Under the new system interculture became the norm; maize was widely spaced in fields and other vegetable crops grown between the maize stalks. These other crops, such as pumpkins and sweet potatoes, filled the spaces between the plants and made weeding unnecessary after the first or second lot of weeds had been removed and the secondary crops had started to develop. This lessened the need for a daily labour input, and it was no longer necessary to hoe throughout the growing season. Intensive labour inputs, including ox draught, were needed at a number of widely spaced points of the agricultural cycle — ploughing and planting the maize seed, perhaps once or twice during the growing phase to remove weeds, and again at harvest time. The smaller household could manage the rest of the day-to-day labour, now concentrated in a single field.

These 'intensive inputs' of labour were provided by forms of mutual assistance or collaborative work in which the members of different households collaborated to plough and plant each other's fields in rotation, and later in the season, to assist with weeding and harvesting. Hunter recognized this development in the 1930s during fieldwork in Pondoland. In large homesteads where there were a number of 'houses', with each wife being responsible for her own field, the women of the homestead as a whole assisted each other in cultivation. In the case of the smaller homesteads, neighbours (who were possibly also kin) assisted each other (1936, 87).

Similar developments have been noted in other parts of Africa (Swindell 1985, 136). In southwestern Nigeria, for example, Painter (1976, 208) attributed new and changing forms of co-operative labour to a pattern of younger males establishing their own independent homesteads earlier than before, leading to a decline in the availability of household labour. In Zimbabwe, too, expanded rural production and the institutionalisation of migrant wage labour had important consequences for the nature of rural households and the social relations between them, including those involving rural labour (Worby 1995). In the Transkei case, state policy and the emerging capitalist economy had rather different effects, the primary one of which was to retard and handicap, rather than to foster and support, the development of rural farmers. In the process, conditions for the emergence of co-operative work were created — a society consisting of small, poor and relatively undifferentiated land-holding homesteads with low cash incomes and no opportunities for cash-cropping (see Moore 1975, 281), but with a determination to maintain homestead production.

In Transkei, co-operative work groups were not necessarily a totally new development. Beinart found no evidence of their existence in earlier times, though they certainly became more important and more widespread, and used for new purposes. Both wealthy and poorer households had access to work-parties, though probably disproportionately in favour of the wealthy (Beinart 1982, 137, 148). However, the existence of a specialized lexicon relating to co-operative work and the beer drinking associated with such labour in Kropf's dictionary, first published before the turn of the century, testifies to the relative antiquity of these institutions.[4]

Other important changes also need to be considered. In pre-colonial times cultivation was almost entirely in the hands of women while men were occupied with the high status task of tending cattle. The basic foodstuff was sour milk rather than grain, though grain and other vegetables, as well as hunting and gathering, made an important contribution to the diet. Women cultivated the soil with a short digging stick, the point of which was flat and a few inches wide. They worked in a squatting position, and the amount of land cultivated was not large. The introduction of the plough and of steel bladed hoes made it possible to cultivate much more land than the earlier techniques of hoe cultivation, but as Heron (1990, 126) points out, it took some 30 years for Xhosa speakers to take to the plough in significant numbers, partly because of the costs involved. By the late nineteenth century, however, ploughs and steel hoes were widespread, the amount of land being cultivated had increased considerably, and large amounts of maize, sorghum and other products were being produced and marketed by Xhosa speakers. By this stage population had grown, land lost to the colonial forces, wild game drastically reduced and the bulk of the Xhosa cattle destroyed in the millenarian movement of 1857 known as the 'cattle-killing'. Grain had become much more important as a food supply, and it was also being marketed to obtain money for taxes and to purchase goods. Land holdings had become larger, fixed and more valuable.

The amount of land under cultivation in Pondoland and in other parts of the Transkei continued to increase considerably in the early 1920s and 1930s, and was reflected in increasing maize yields (see appendix in Beinart 1982). For the smaller homestead this meant that acquiring implements and building up a herd of cattle were vital, and the primary

---

4 Some of these terms *are amandwandwa, ukunyenyetha*, and *umtabata* (see below).

means of achieving these objectives was through migrant labour. Ox-draught, however, put much of the important agricultural work in the hands of men. Women were forbidden to work with cattle, but their total labour input probably increased due to the much larger amounts of land being cultivated, since they were still responsible for tasks such as weeding and harvesting. Ox-draught may well have contributed towards entrenching patriarchy and lowering the status of women, however, because it deprived them of sole charge of garden produce and placed cultivation under the control of men. As Heron (1990, 131) puts it,

> The impact of the plough on 'traditional' Xhosa society was vast. It created new wealth differentials while at the same time reinforcing old ones; altered the division of labour; increased men's power over women; changed the ecology of the area because of greater areas under cultivation; pushed people towards a more sedentary way of life; and brought cattle into agriculture. This also meant that the economic importance of cattle increased. They were now essential for use as draught and they were also a source of manure — a replacement for ash in the old slash and burn system.

At the same time, labour co-operation between homesteads became vital to solve the labour shortages created by migrancy and declining homestead size. In this way, despite the difficulties and contradictions, and the fact that some households did not manage to cope, 'the evidence suggests that total and perhaps even *per capita* input may have continued to increase during the decades when mass migrancy became institutionalised' (Beinart 1982, 100).

Maintaining homestead production in Transkei also needs to be understood in the context of the development of racist and segregationist policies from the time of annexation through to the development of apartheid. These deprived blacks of land and freedoms, prohibited them from urbanizing, and forced them to attempt to conserve and maintain household production as far as possible. In Pondoland, households 'had been blocked from access to external markets for their grain and stock and were intent on protecting their natural resources for communal purposes' (*ibid.*). Of course, this was eroded over the years, especially by increasing land shortage and betterment schemes, but in areas where betterment was successfully resisted and where land availability and lack of population pressure made it possible, homestead production remained important, though after about 1940 it was seldom more than a supplement to wage labour for the majority of homesteads. The question of maintaining homestead production as both practise and ideology remains crucial to understanding co-operative work and beer drinking in Transkei, and will be revisited in the conclusion.

As Beinart points out, in Pondoland the new style of agricultural production could not continue to meet household food needs for long after the 1930s, as population pressure and land scarcity grew, and as lands became infertile due to over use and erosion. These problems were partly resolved in Gatyana district, and adjoining areas of the Transkei, by yet another adaptation, the switch from field to garden cultivation, as outline above in chapter two. The persistence of homestead production in South Africa was, as historians have commented, combined with a flexibility which allowed it to survive in diverse circumstances (Marks and Atmore 1980, 10-11) and one of the points of this analysis is to illustrate that this flexibility has persisted for far longer than is commonly thought, at least in certain areas.

The development of beer drinks

It is important to point out that there are parallels between the historical development of co-operative work as outlined above and the development of beer drinking in the Transkei. Early observers of Xhosa-speaking peoples made very little mention of beer, let alone of beer-parties. Kay (1833), who made careful observations of Xhosa diet and ceremonies, claimed that Xhosa people were not at all familiar with beer produced from grain, though in Pondoland he noticed that people made it in large quantities and used it frequently. Gardiner, travelling in Pondoland at roughly the same time, made a similar observation, but pointed out that the prominence of beer drinking was due to the scarcity of the staple food, milk, as a result of cattle losses to Zulu raiders (1836, 266). Beer had taken the place of milk as a basic food.

There are indications that beer became much more frequently made and consumed in Transkei in the 1870s and 1880s than it had been fifty years earlier. Evidence given to the 1883 Commission on Native Law and Customs indicated that beer had become common, and had replaced the use of milk at feasts. This was a time when cattle were scarce due to a series of diseases that afflicted the Transkei in the 1870s. Later, diseases such as rinderpest, which swept through the country in 1897, and East Coast fever (1912-13), destroyed as much as 80% of cattle in some areas (Beinart 1992). Beinart (1982) points out that when maize was plentiful in Pondoland, the incidence of beer drinking increased, and Hunter (1936, 357) says that old people in Pondoland were 'emphatic' that beer was being consumed in larger quantities than when they were young. In Bomvanaland, situated between Gatyana to the south and Pondoland to the north, the increased frequency of beer drinking in the 1920s caused great concern among older people (Cook 1931). Clearly, the larger amounts of grain which became available in the first decades of the twentieth century, as a result of the factors outlined earlier, and combined with increasing population and a shortage of milk, led to increased beer-drinking.

The reports of the United Transkeian Territories General Council (the 'Bunga'), in the period between 1900 and 1930, confirm this trend and also point to the development of new forms of beer drinking ceremonies,[5] which cannot be explained simply in terms of a shortage of livestock and milk, even though in some cases these were beer-drinks which replaced earlier forms of ritual in which a goat was killed for the ancestors — the *umsindleko* beer drink for a returned migrant is one of these, and the beer drink to mark the establishment of a new homestead is another (McAllister 1985, 1993a). The implications of substituting a beer drink for a ritual slaughtering are significant, in that they coincide with and give symbolic and supernatural support for a change in the nature of the worshipping group from one based primarily on kinship to one in which neighbourhood is very important. These substitutions developed in tandem with a changing mode of production, where the reliance on kinship-

---

[5] For a detailed review of this evidence see McAllister (1993).

based labour within the extended family household was largely replaced with a reliance on neighbours and co-operative work based on spatial rather than kinship principles.

Kinship and neighbourhood organisation

We are now in a position to turn to contemporary data on work parties in Transkei, to resolve some of the differences in interpretation referred to above (e.g. the moral basis on which Xhosa people join and organise work parties) and to bring beer more fully into the picture. In what follows I present data on work parties and beer drinking in Shixini. Before embarking on this it is necessary to provide a basic outline of two important principles of social organisation in Shixini, kinship and neighbourhood.

Xhosa-speakers are patrilineal, and each person belongs to a patronymic clan, members of which are widely dispersed over the entire country and beyond. Clan members claim common descent but cannot demonstrate their blood links. Like clans, lineages do not exist as corporate groups and are widely dispersed geographically, having no corporate functions. This applies even to lineage segments, members of which have to be called from afar to attend rituals, this being the only occasion on which they come together. Lineages and clans have no 'on the ground' existence (Hammond-Tooke 1984). The effective kinship group on the ground is a small 'agnatic cluster', consisting of perhaps a group of brothers and their children, or a father and two or three sons, each with their own homesteads (*ibid.*). Such clusters frequently inhabit nearby homesteads in a single local area. Close co-operative relations, arising out of the historical process outlined above and based on an ideology of kin group corporateness, may exist between members of the 'agnatic cluster', as it may between other categories of kin who live in a common locality, but seldom to the exclusion of others (non-kin) from such relationships.

As mentioned in chapter one, Shixini is divided into sub-wards. These consist of up to 130 homesteads each, and have a distinct sense of historical unity and solidarity based on territory and the historical establishment of the sub-ward under a particular founding leader who was usually a kinsman to the chief. Each sub-ward is in turn divided into two or three territorial sections containing from 10 to 40 homesteads each. A section may be associated with a particular clan or lineage segment, or with an agnatic cluster, but this is not always the case, and seldom do all the members of such a segment live there. Sections always include members of a variety of different clans, and some of these members may be related cognatically or affinally.

It is at the level of the sub-ward and its sections that virtually all, everyday economic and political activity takes place. For example, control of land (homestead sites, grazing, arable allotments, etc.) is exercised at the section level, and decisions and allocations ratified by the sub-ward as a collectivity under the sub-headman, and then by the chief. Sections are also important in ritual and ceremonial matters, as detailed below, and in law. Senior male members of the section constitute a court for the purpose of resolving local disputes and hearing cases involving minor local misdemeanours, while more serious matters go to the sub-ward moot, from where they may be referred to the Tribal Authority. Sections are, in effect, closely knit neighbourhood groups, which play a vital role in many aspects of social and economic life, including the organisation of co-operative work and the conduct of beer drinks.

The organisation of agricultural labour

One of the most striking features of life in Shixini (despite and because of a heavy reliance on labour migration) is the dense network of relationships between homesteads based on both neighbourhood and kinship, involving collaboration and interaction in a variety of agricultural and other activities. As alluded to in the section on maize yields, it is largely due to this that homestead production is possible. Kin and neighbours join their animals together to facilitate herding arrangements; they form a variety of work parties to overcome shortages of labour, implements and oxen; and they assist each other in a variety of small but significant ways on a day-to-day basis. Labour combinations are also a means of re-distribution, and to many households providing labour for others is an important (though always partial) means of securing subsistence, as well as reciprocal labour at some time in the future.

In what follows, the various types of labour combinations are outlined, and particular attention paid to the beer drinks that frequently mark their conclusion. This is an important aspect of the larger work parties, because it is here that participation in the work is dramatized and the experience of labour placed within a wider social and ideological context. An examination of this is essential for a proper understanding of the organization of work parties and, more generally, of social life in Shixini as a whole. However, it has to be conducted in the context of a wider understanding of beer drinking in Shixini, which is also associated with a variety of activities other than work. After a consideration of work parties, therefore, we turn to a general account of beer drinking in Shixini, before bringing beer and work together again in the following section.

Work parties and their associated beer drinks also need to be examined against the background of other kinds of labour in Shixini, to which they are related in various ways. In many cases, extra-domestic labour is *ad hoc* and informal and does not take the form of a work-party. Close kin and neighbours often assist each other informally with tasks without any material goods changing hands or any expectation of direct reciprocity. A mother may ask her daughter from a nearby sub-ward to come and provide assistance in the garden; the women of neighbouring homesteads may assist each other on a casual basis without giving or expecting any material reward. Relationships between homesteads of a patron-client type (which may coincide with agnatic links), similarly, may involve labour service on the part of the poorer client in recognition of the diverse and frequent material and political assistance received from the patron.

Sometimes members of a small number of homesteads (from two to four), agree to help each other on a reciprocal basis whenever necessary. Such groups are small and not 'work parties' as such. They are usually composed of close neighbours, sometimes consisting of a number of agnatic or other kin. Their activities are inconspicuous, taking place without public announcement or public involvement. Unlike most work parties, no beer is provided at the conclusion of the work, which helps explain why the work that they do goes more or less unnoticed.

Daily tasks in and around a homestead are performed by its resident members. These include work such as cleaning and cooking, herding and attending to livestock and land, repairing huts and byres, collecting firewood or dried dung for fuel, weeding, harvesting, drawing water from the nearby stream or spring, walking to the trading store for necessities such as maize meal, tea or sugar, and so on. Certain tasks are associated particularly with women and girls, such as tending small children, stamping and grinding maize, brewing beer, re-dunging walls and floors, cutting grass, or feeding pigs and chickens. Other tasks are associated with men and boys, including tending livestock such as goats and cattle, milking, herding, slaughtering animals for food, felling poles for fencing or byres, fencing, building, ploughing, and the like (see Liebenberg, 1997). Many tasks, such as weeding and harvesting, may be done by both men and women, and women may perform tasks strictly associated with men (usually those relating to cattle), and vice versa, if there is no alternative. Some of these are seasonal tasks, linked to the cultivation of maize and other garden crops; others are sporadic and infrequent (e.g. hut-building); yet others are performed daily.

Some of the chores associated with daily work may also be done collectively, either with the help of a small group of neighbours, as described above, or through recruiting a larger and more formal work party, to which we will turn presently. Those collectively performed tasks which involve female labour include cutting thatching grass for a new hut or the repair of an old one, collecting firewood (for example, if a large amount is required for some ritual or festive occasion), making mud-bricks and harvesting. Male labour is required for collectively cutting poles, constructing huts, byres or fences, and for ploughing and planting (which are usually done together) (Heron 1990). Work parties performing tasks such as hoeing commonly involve both male and female labour. Hoeing to clear weeds from a field or garden is in fact the most common form of collective labour in Shixini, along with ploughing and planting, with harvesting, cutting poles and cutting thatching grass not far behind these in frequency.

A more formal but relatively uncommon way of organizing neighbourhood labour helps to throw some of the more general principles into relief. In Folokhwe sub-ward in February 1998, there was at least one formal hoeing group, referred to as an *umbutho* (literally a 'regiment'). It came into being in 1996 when a group of eight adult women decided to hoe and perform other kinds of work for each other on a reciprocal basis. Hoeing and other tasks (such as making bricks, cutting thatching grass, collecting firewood) are performed on the basis of strict reciprocity rather than in return for some commodity, although the person requesting the others to work must provide tea and bread in the course of the morning, and food (e.g. samp or *marhewu*, a drink made from fermented maize-meal) for the workers once the job has been done. Reciprocity is generalized within the group; that is, hoeing may be reciprocated with another form of work if a member is unable to come and hoe on an appointed day. The existence of the group does not preclude a household from making other arrangements to do any particular task, including hoeing, nor does it prevent individual members of the group from entering into other kinds of labour arrangements with other households or with each other.

Many homesteads do not enter formal reciprocal alliances of the above kinds, but nevertheless co-operate extensively with neighbouring homesteads in a similar way by means of small work-parties based on requests made for assistance. A small work group of this kind is called an *isicelo* (a 'request', from *ukucela*, 'to ask for something'). It is similar in form to the *ilima* described by Kuckertz (1985), but smaller in scale. The status of the work as response to a request for help is encapsulated in the name. Some informants refer to an *isicelo* as an *ilima*, others say that an *ilima* work group is one that performs tasks associated with cultivation (*ukulima*), a usage that concurs with Kropf's definition of the term (1915, 216), and yet others say that an *ilima* is a sort of general purpose word referring to any kind of work group (which is very much how Hunter (1936, 88-92) uses the term). Anthropologists might wish to define these things with precision and certainty, but in everyday life they are often characterised by vagueness and ambiguity!

To hold an *isicelo*, a homestead head goes herself or sends someone else to ask neighbours to come and assist with a particular task, also notifying them of the nature of the reward. The latter may be a small amount of maize-beer, but more frequently it is an alcoholic drink called *mangumba* (*jabulani* in other parts of Transkei), made overnight from shop-bought ingredients.[6] Or the reward may consist of some other commodity or combination of commodities such as paraffin oil, meat, sugar, maize, samp and bean soup, or *marhewu* (a fermented maize-meal drink). The reward offered is not regarded as a payment, but as a necessary incentive provided 'so that they do not get hungry, since they have come to help'. When Nophatile organised the thatching of her son's hut in July 1997, she held two *isicelo* work parties for cutting grass. For the first she provided *mangumba* (4 beakers) as well as five bottles of Castle Stout for the four women who came to help her. On the second occasion a similar amount of *mangumba* and stout was provided for the three who came to assist. When Ntanyongo needed help in removing some very large weeds from his garden, he called on three neighbouring women to assist and brewed *mangumba*. In the event, rain made it impossible to do the work but the *mangumba* was consumed nevertheless. The women, all from Ntanyongo's sub-ward section, returned a few days later, after it had stopped raining, to complete the job. This time Ntanyongo had prepared a small quantity of home brewed maize beer (*umqombothi*). The presence of beer meant that men from immediately adjoining homesteads could be (and morally, *had* to be) invited to partake of it, one beaker being set aside for them. There were two beakers for the women, who were also fed with *marhewu* and with samp and beans.

Not every close neighbour is necessarily asked to participate in such a work group. As in the Mpondo case (Kuckertz 1985), the head approaches people whom he/she is on good terms with and who will be expected to help. Failure to heed the request to assist in such a small *isicelo* without a very good reason would be a breach of friendship and neighbourliness. Nophindile recognized this quite clearly. She had decided, in February

---

6 Strictly speaking, *mangumba* is not 'beer', which is brewed from maize and called *umqombothi*.

1998, to provide 'paraffin for hoeing' and she sent word asking people to come and hoe. Those who were asked were all people she had a close relationship with; they were close neighbours within the sub-ward section, and some were also close kin. Nophindile had in the recent past assisted some of them with weeding. All those who were asked to help came along to work on the day. *Nophindile* pointed out that this was *ukuncedana* (mutual helpfulness) and that she would have been very disappointed if any of those asked to help had not agreed to work. But she also pointed out that such work had to be rewarded; 'the person I hoed for will be expected to hoe for me, but this does not mean that she should be asked to hoe without reward!' However, a young women whom we spoke to said that she herself often participated in large work parties even though as a Zionist church member she did not drink beer: 'You do it because of *ukuncedisana* ('helping each other'), even if you do not drink.' She did receive tea and something to eat at these events, but said that 'I did not go for the food, only to help.' However, this ideology of mutual helpfulness needs to recognized for what it is, a moral norm in terms of which, in practice, assistance is given to people with whom one has some kind of actual relationship based on friendship, neighbourhood, kinship, or obligation.

A somewhat different way of securing labour through the provision of commodities is to make the commodity available, and to invite people of the neighbourhood to partake of it in return for labour, either immediately for a specific task, or at a later date. Passing by the homestead of Kanana one hot February day, we observed three women hoeing in the smaller garden (the *mvelidinga*). We stopped and asked to speak to *umama wekhaya* (the 'mother of the home'). One of the women came over — it was Noncedile — and after the preliminary greetings and inquiries we asked her who the women helping her were. She replied that one was the wife of a neighbour and junior agnate of Kanana, her late husband, and she described her as *umolokazan' am* ('my [classificatory] daughter in law'). The other was also described from the male viewpoint, as the daughter-in-law of Dhyubeni, not a close relative, from another part of the sub-ward. One could easily have come to the conclusion that this was a small work group based on kinship principles. But although Noncedile explained who the women were in terms of a kinship idiom, she then also added, almost as an afterthought, that they were 'working for meat'. Asked to explain further, she told us that she had killed two pigs, referring to them as *iihagu zokuhlakula* ('pigs for hoeing'). Those who accepted meat from these pigs had done so on the understanding that they would 'pay' for it with their labour at hoeing time. It emerged that there were well-established conventions relating the amount of meat taken to the number of days of labour payment (a 'day' here refers to the period between about 6 a.m. and midday). A joint (back- or fore-leg) required three days work; a side, two days. However, this also varied according to the size of the pig. In this case, six people, not all of them kin to Noncedile, had accepted meat and had promised to hoe in return, and the two in the field with Noncedile at the time were the last to fulfill their obligations.

An advantage of securing assistance in this way is that the head can call in the labour debts when he/she feels it is needed, and need not call all of those who owe at the same time. It is more usual, in fact, for one or two of the debtors at a time to be asked to come and work, alongside members of the host homestead. If one goes around and invites specific people to come and work on an appointed day, announcing that a pig will be slaughtered, one does not know how many people will turn up, which increases the risk that the investment will not be a good one.

The difference between this form of work and an *isicelo*, based on a request for help, is that no individual neighbour is personally asked to volunteer his/her labour. Instead, an invitation is made, those who are approached or who come to hear of it indirectly and who feel like doing so will take it up. However, taking up an offer of this kind is more than simply accepting a payment for labour service, it is also a friendly neighbourly act which people expect from those whom they have a close relationship with and which they regard as deserving of reciprocation at a later date. As with *isicelo*, the basis of the relationship is clearly more than simply material, but the material element is necessary for the transaction. Thus long-term reciprocity on the basis of neighbourhood, kinship or friendship co-exists with short-term reciprocity — the exchange of labour for goods.

However, there is possibly a qualitative difference between this form of labour exchange and responding to an *isicelo*, the reward for which is much smaller in material terms, and which carries a correspondingly stronger reciprocal obligation on the part of the homestead which benefits from the labour. This is indicated by the fact that there are circumstances in which the exchange of paraffin or pork (or some other item) for labour appears to end the transaction and to carry no reciprocal labour obligation. These circumstances concern how and by whom the transaction is initiated: For example, in one case a number of women who assisted with the hoeing of Nomawele's garden had come to offer their labour in return for maize, which they knew Nomawele had recently purchased. Nonzame, one of these women, had encountered Nomawele in the taxi *en route* from Willowvale town, where she had been to make the purchase. Nonzame, who lives a good distance away in another sub-ward, came over the next day to ask if she could have some of the maize in exchange for hoeing. The others were neighbours and kin, but Nomawele felt that kinship and good neighbourliness had nothing to do with the exchange; they came to work in order to be given maize.

In another case, three young women from Jotelo sub-ward (also in Shixini) came to Folokhwe sub-ward and went from homestead to homestead asking for hoeing work in exchange for paraffin. One of the women who took up their offer spoke of their work as *isintu* ('human kindness') but did not feel that she would be obliged to help them in future due to the fact that it was they who had come and offered their labour in exchange for a payment in kind. Clearly, not all cases of labour assistance follow the same pattern or involve the same kinds of rights and obligations. The use of the term *isintu* is significant, however, and indicates the presence of a moral element; it also means 'human-ness' in the sense of 'human custom' or culture, and carries a connotation of moral rightness. It is similar to the well-known concept *ubuntu*, the abstract quality of 'humanity', of being a proper, upright human being.

This was brought home to us very forcibly when we noticed the mother of the wealthiest man in Folokhwe hoeing in the garden of someone who was neither a relative nor a close neighbour. Why, we asked the female homestead head, was this person hoeing in her garden? 'For tobacco' was the response. Apparently, she had come to ask if she could hoe in exchange for tobacco. In economic terms it made little sense — not only was her son very wealthy by local standards, but she herself was a diviner and income-earner in her own right with ample ability to pay cash for the tobacco she wanted. On further inquiry it emerged that the homestead had, at an earlier point, also enjoyed the labour of four other women who had hoed, this time for paraffin. All four of these women were close neighbours and kin, and all four came from homesteads that were not by any means poor — one was from another very wealthy homestead. None of them had to work because they needed paraffin and could not afford to buy it. They worked because they were asked to do so and, in the case of the one wanting tobacco, because labouring for others is socially approved and valued, and because of the expectation of labour reciprocity at a later date. What these women had in common apart from a lack of poverty, was that they were from homesteads that were all active agriculturally, and which would need to host work-parties for weeding and harvesting from time to time.

On rare occasions, workers are employed for a cash wage, but this is well below the cash value of the goods offered in the exchanges of commodities for labour. A day's work is worth R5 if paid for in cash, but R9 if paid in paraffin (5 bottles or 3,75 litres per day), R10 in tobacco (approx. 700 grams), R12 in dried beans (1 small tin-full — approx. 1 kg), R12,50 in pork, R12,80 in maize (2 billy-cans full, or 4 kg.), and R17,35 in sugar (4 small tins-full or 5 kg per day).

It is difficult to determine the logic of this. To some extent what is offered in exchange for labour depends on what is available in the recruiting homestead, whether in terms of local agricultural produce or goods brought home by a returning migrant. But the goods offered in exchange for labour are consistently some two to three times the value of the conventional cash wage. A possible explanation for this revolves around the question of reciprocity. A cash transaction carries no further obligations on the part of either employer or worker, but as we have seen above, exchanges of goods for labour initiated by the recruiting homestead are governed by a moral norm which stresses labour reciprocity, on the one hand, and a material reward for labour, on the other. At first glance at the Rand values listed above, and from the 'purely economic' point of view, one would expect those who want labour to offer cash, with those offering labour to demand sugar. However, women claim that out of these various rewards, it is cash and paraffin that work best in securing workers. What this suggests is a premium on hard cash, since of the various commodities used, it is only paraffin and sugar which can not be produced within homesteads, and of these two items, one can possibly do without sugar but not without paraffin if this is the primary source of household energy. People say that it is 'too expensive' to hire workers for R5 per day, and that they provide maize, beans, or other commodities instead. On the other hand, many households do have to purchase maize and beans, since the quantities of these items, which they grow in their gardens, are often insufficient for household needs. But looked at in terms of reciprocity, in the form of both labour and reward for labour, both parties can assume that there will be a balancing out in the medium or long term, at least, and therefore it is more 'profitable' to offer a commodity such as paraffin in return for labour than it is to offer a cash wage.

The other disadvantage of engaging wage labour is that it signals that the transaction is a western-style, commercial one, not associated with reciprocity. In this it is contrary to the general ethic of mutual helpfulness and the values associated with good neighbourliness. In non-wage transactions in which labour is rewarded with some kind of commodity Shixini people are at pains to stress that the reward is not a payment. This is symbolised at work parties for hoeing, as Heron has pointed out, when the workers deliberately leave a patch of the garden or field un-weeded for the homestead to do itself. If the beer or other reward provided was a payment, they say, the homestead head would be entitled to demand that they complete the job. Leaving a patch un-weeded symbolises the fact that they are not hired labourers and that they were 'just helping' (Heron 1990, 108).

Payment in cash or kind for hoeing is by no means a new thing in Shixini. All the people we asked about this were able to cite earlier cash rates for a day's labour (various informants recalling different periods in the past mentioned rates of nine pence per day, one shilling a day, two shillings and sixpence a day, etc.). Ngcinibandla, about 65 years old in 1996, said that killing a pig for hoeing had been common when he was a boy, and that the use of commodities such as paraffin and sugar was a long established custom. What was new, he said, was giving maize for hoeing, because in the old days everybody had maize. This was confirmed by a younger woman of about 35 years of age. But two elderly women said that hoeing for maize had been practiced when they were girls, and that they used the maize to brew beer for sale, using the proceeds to buy things such as scarves for themselves. At the time they had maize at their parents' homes, they said, but did not feel free to use it in this manner, preferring to work for their own.

An individual may use a variety of strategies, including wage labour, to ensure that his/her garden is adequately hoed. For example, in the 1996/1997 season, the hoeing in Masilimane's large garden was done partly by the two adult women of the homestead — his wife and his FBD, partly by women who had asked for seed earlier in the season and who had agreed to hoe in return, partly through hiring labour at a cash wage of R5 per day, and partly through killing a 'pig for hoeing'. Masilimane's wife is a member of the *umbutho* mentioned above, and this group, too, was responsible for some of the hoeing done. In addition, Masilimane requested help from an *umtshana*, a FBDD who lives nearby. She gave her labour free of charge because, as Masilimane said, 'she always gets maize from me when she needs it, so she is not paid for hoeing.' He acknowledged that the basis of this was *ukuncedisana*, 'helping each other', but when questioned about those to whom he had paid cash he said that it was the payment that was important, not *ukuncedisana*. 'They may need money', he said, 'and their own maize may be very immature and not yet in need of hoeing.' On reflection, however, he added that even with cash payment for hoeing, there was an expectation of reciprocity: 'There is *isintu* involved; if people notice that you do not help others they will not help you, they will leave you alone with your paraffin or your maize [i.e. being offered for hoeing].' As an example he cited the case of an elderly woman who lived nearby, who never attended other people's ritual events and never ate meat at any homestead other than her own. When she had recently held a small ritual for her grandson, at which a goat was killed, the women who attended refused to eat any of the meat placed before them as an indication of their disapproval of the woman's unsociable behaviour.

Masilimane's garden is, however, exceptionally large (as described in chapter three), and he takes a great deal of care to ensure that he has good yields — removing the weeds at least twice during the growing season. Many homesteads with smaller gardens and enough labour do not have to engage any assistance with hoeing, especially if they are happy to remove weeds only once and therefore accept the lower yield that this brings.

Agricultural companies and work groups

One of the more important forms of co-operative work in Shixini is what is colloquially known as the 'agricultural company' (*inkampani yokulima*). This consists of from four to twelve homesteads, which combine their labour, implements and draught cattle to perform agricultural tasks such as ploughing, planting and, to a lesser extent, harvesting.[7] They are extremely important groups, and without them agricultural activity in Shixini would be difficult to sustain. Most homesteads belong to one of these companies, though there are a few who do not, either because they do no cultivation, or because they prefer to work independently and are wealthy enough to do so. As with the *isicelo*, these groups operate on the basis of notions of reciprocity and co-cooperativeness. In one sense the reciprocal element is both short-term and quite strictly adhered to, in that

7 It may be that these are similar to what Kuckertz describes as 'reciprocally co-operative groups' in Pondoland, which help one another 'at harvest time or with a span of oxen.' Kuckertz considers them to be 'relatively unimportant' (1985: 118), and excludes them from his analysis.

work such as ploughing is done for each member in turn every season, and a member who fails to contribute as expected, for no good reason, can be forced out of the group. In another sense, however, reciprocity is longer-term, in that resource imbalances are recognized and may have to be tolerated within the company for many years. No beer is brewed when company members come together to work at the request of a member, though their contribution to the homestead is recognized when it brews 'beer for harvest' later in the season (see below). If they work on behalf of a non-member, however, the latter must brew for them.

Recruitment to these companies is based on a variety of factors. Some companies seem to be based primarily on agnation, with the members of a local agnatic cluster belonging to a single company. Such companies may be conceived of in terms of kinship solidarity and the obligation of kin to help each other. Most companies, however, include members of a variety of agnatic clusters and clans within a common locality. The members of an agricultural company, whether they are agnates or not, are almost always drawn from a single sub-ward section or sub-section. Occasionally they include a member from an adjoining sub-ward, but this is rare and usually due to the voluntary re-settlement of a member (or his son, since membership is partly hereditary). One could argue, therefore, that the effective basis of recruitment to companies is locality rather than agnation, since agnates within a common territory do not necessarily plough together, and agnates living in different areas cannot do so, for practical reasons. This would, however, oversimplify the matter. The kinship component needs to be considered carefully.

In Nompha, the Shixini sub-ward studied by Heron (1990), 70% of all members of agricultural companies were agnatically related, and a further 22% were related cognatically or affinally. In the 11 companies in the sub-ward however, only three consisted solely of agnates. Two of these each consisted of only two homesteads, the third of five homesteads. In only one case did all the members of an agnatic cluster belong to one company. This was a group of two bothers and a deceased brother's wife who had come to live in Nompha relatively recently (*ibid.,* 145). They formed a company together with a sister and a wife's sister.

In Folohkwe sub-ward in 1977 there were 10 companies, to which 61 homesteads (77% of the total) belonged. The majority of members of agricultural companies were kin to each other, but only 25 (41 %) were of the same agnatic cluster as the company's head. Eight of the ten companies included non-kin as members, and no company consisted only of close agnates (or of agnates and the widows of deceased agnates). Classificatory kin, including people of the same clan but with no demonstrable blood link, and non-kin, accounted for 28% of the members. In six of the ten companies the majority of members belonged to the same agnatic cluster and in the other four the number of non-kin was greater than or equal to the number of agnatically related kin in the company. As in Nompha, virtually all members of every company were from the same sub-ward section, and only two came from outside the sub-ward.

Heron also points out that the companies are associated with an ideology of kinship. For one thing, they are referred to as *amafelandawonye* — literally, 'those who die together', or 'die-hards', indicating that the bond between them is a long term one and is based on ties which cannot be broken. The companies have to be formed on kinship principles, Heron argues, because this provides the moral basis for the tolerance of long-term imbalances in resource contributions. It also reduces risk. In a situation where many factors can affect the outcome of agricultural effort, being able to rely on the other members of the company is vital, and kinship provides this reliability. A migrant worker away at work, for example, can be assured that his kin will plough his land for him in his absence. Heron also points out that certain members of the companies contribute very little by way of labour or other resources, and that their dependence on the other members is best understood in terms of a morality of kinship which tolerates long-term imbalances and stresses the need to support kin who are in need. Yet Heron's case studies (1990, 133-155) of companies forming and splitting illustrates that companies do split when they grow too large (sometimes along lineage segment lines, sometimes not) or, quite frequently, as a result of friction between members. Kinship is no guarantee that the members will not fight, nor is it a guarantee that they will honour their obligations.

However, data from other parts of Shixini indicates that the dependent and largely non-contributing members are not necessarily kin to the majority of company members, and that this obligation is a generalized one not specifically linked to kinship. In Folokhwe a number of people in this category may be identified, who might be called 'affiliated' (as distinct from full) members of the companies. Sixteen of the 61 homesteads belonging to companies fell into this category. Of these, ten were not related to the head of the company, three were agnatically related to the head, and four were related to the head in other ways (McAllister 1979, 268-269).

Participation in other kinds of non-kinship based work parties, too, may be marked by long-term imbalances, as Heron recognizes (*ibid.*, 110). The term 'die-hards' (*amafelandawonye*) can also be used in a non-kinship sense, for example as a general term for rural co-operatives (Beinart 1987). So perhaps these companies are best characterized as being of diverse composition; some consist only of kin, in which case membership and activities are characterized by a kinship ideology; in other cases they consist of members of a variety of agnatic groups and include non-kin. In this case, the ethic of mutual helpfulness need not rest on an idiom of kinship but is more general and based on good-neighbourliness (*ubumelwane*). In both cases, membership is recruited from the immediate sub-ward, usually exclusively from the sub-ward section, and in this sense they are neighbourhood groups even in those cases where the majority of members are close kin.

The companies are relatively enduring — their membership remains stable from year to year, though changes occur when companies split due to becoming too large or for some other reason. Occasionally, new members are accepted into a company or an existing member leaves to join another. Not all companies have acknowledged leaders but are based instead on consensus among the majority of members, especially those who make significant inputs in terms of cattle and labour (which in many cases includes all the members). In other cases a prominent or wealthy man is seen as the leader of the company, and the companies in a sense form around him. It is in such companies that 'affiliated' members are

Photo 2
A small work party clearing weeds from a maize field

most likely to be found; people who have little to contribute but their political loyalty to the company's head plus some labour in times of weeding and harvest.

In this sense no two companies are alike, though at the normative level they are explained in terms of the moral norms of kinship and 'mutual helpfulness' (*ukuncedana*) and of helping those who are in need. So although recruitment takes place, in statistical terms, from within sub-wards and sections, the actual networks of co-operation in ploughing groups reflect a variety of other factors, based on mutual interest, individual resources, friendship, individual ability, past experience of working together, reputation, political allegiance, and so on (see Holy 1977). The moral norm that kin ought to plough together, for example, is modified in a variety of ways by practical and social considerations, such as the actual distribution of labour, oxen and implements, or by relationships of close friendship or antagonism between kin. Relationships within the group, for example, depend in part on the resource contributions they make rather than on their kinship relations, as some of Heron's case-studies show. In entering into arrangements of this kind, individuals are not simply acting in terms of norms associated with kinship and neighbourhood, but also pursuing the practical strategy of developing their homesteads, a topic which is of crucial importance in understanding both work and the beer drinking associated with it, as we shall see below.

The need for agricultural companies arises for a number of reasons, both ecological and material. The optimum time for ploughing and planting is limited to a few weeks at the beginning of each season. Heron (1990) has shown that once allowance is made for all the days on which this work is not possible due to weather and soil conditions, or not allowed for by cultural ones, there remain relatively few days on which to get the ploughing and planting done in the three-month period October-December when this work is normally completed. In the 1988 agricultural season, for example, Heron calculated that there were only 26 days available for ploughing. Naturally, such a figure will vary from year to year. Shortages of oxen, implements and labour make it necessary for people to pool their resources and to plough and plant each other's land in rotation if cultivation is to take place at all in most gardens. Only one quarter of the homesteads which cultivate own ploughs, 13% own mechanical planters, and less than 10% have cultivators. Livestock holdings are detailed in chapter three, above, and it is clear that most homesteads with cattle have to combine their oxen with those from other homesteads to make up a ploughing team of six. It is on this basis that the companies are formed, and on which they split, either because they grow too large to rotate the work effectively, or because of dissatisfaction or disagreement about the order of work in relation to the relative contributions made (*ibid.*). This is a crucial point in the agricultural season, and the importance of optimal ploughing and planting creates a certain amount of pressure and tension (*ibid.;* Gulliver 1971).[8] This illustrates again that interest is an important aspect of the membership and operation of companies.

The order of ploughing and planting in companies turns largely on the resource contributions that are made by the different members. In companies with a clear leader or head, this position is generally associated with a large resource contribution, and the head and other prominent, contributing members of the company, have first choice when it comes to having their land cultivated. This usually, but not invariably, means that they get their lands cultivated at the optimum time. In many companies, however, resource contributions may be more or less equivalent, and the members may take it in turns to have their land ploughed first and last. Members who make very little contribution to the company's resources usually have little choice about the timing of their cultivation, and must usually wait until the other members have done their own land. In practice there are a large number of variations in the ways in which the companies work for their members, and the conventions, which are established by the company, may in any case change from year to year through negotiation or expediency.

Herding groups are usually composed of neighbouring homesteads that also belong to the same agricultural company (in small companies the

[8] Conflict and tension is perhaps more likely to occur where there is a lack of formal cultivating groups and homesteads have to compete with each other for labour, as among the hoe-cultivating Ndenduli (Gulliver 1971).

overlap with the herding group may be complete). These homesteads pool their livestock for herding purposes, and for dipping, in the case of cattle, thereby economizing on labour.[9] The animals leave their individual byres each day and are herded to the grazing areas, where the herdsman (often a boy) keeps watch on them until it is time to return again in the afternoon. This is particularly important in the growing season, when a few cows astray in someone's garden can wreck the maize crop in a very short space of time, and cause ill-feeling between neighbours.[10]

Umgqibelo and other labour combinations

Agricultural companies or parts thereof may assist each other on an informal basis, and sometimes two companies which were once united have a standing agreement to call on each other in times of need. Alternatively, a homestead may brew beer and ask a number of companies to assist it with cultivation. This is called an *umgqibelo* group, and it is usually composed of companies drawn from the same sub-ward section, almost invariably from within the sub-ward. The homestead head making the request may or may not be a member of one of the companies involved. Usually the task is to plough and plant the homestead's field or garden in one day of intensive work, thus enabling the homestead to take advantage of favourable weather and soil conditions, and to get the work done at a time which will help to ensure a good harvest. The term is derived from *ukugqiba,* 'to complete' or 'to finish', and in non-work party contexts

---

9 Five sub-wards in the lower half of Shixini share a single dipping tank, and cattle have to be driven to the dipping area and back on dipping days.

10 Many of these details on co-operative labour in Gatyana, collected between 1976 and 1998, are surprisingly like those recorded for the Mpondo by Hunter (1936), based on fieldwork in the 1930s, and for the Xhosa and Mfengu people who were the subject of the Keiskammahoek rural survey, fieldwork for which was conducted in the late 1940s (Wilson et al. 1952). In both cases work parties were based primarily on locality, participation implied reciprocity of some kind, and reliance on them was extensive. Kin would be invited to attend if they lived nearby, but the obligation to assist kin was no greater than that to assist neighbours. Rotational ploughing and planting among an enduring group of neighbouring households was common in both Keiskammahoek and Pondoland, and other work parties were organised for cutting grass, collecting firewood, etc. The reward was beer for the larger groups, in which non-workers shared, but meat or other food was sometimes provided for an ordinary work party *(ilima),* especially in Christian households where beer was prohibited.

umgqibelo means Saturday, which ends the working week. Heron (1990) points out that this work is frequently planned for a Saturday because in the cultivation season this is the only day on which companies are not busy, though weather and other factors may make it necessary to do the work on any available day.

Between some small companies (usually only two) there is an agreement to form *umgqibelo* on a reciprocal basis on behalf of any individual member of the companies involved. In such cases beer is not brewed. Should the *umgqibelo* require a third company, however, or be called by someone not a member of the companies, or involve companies that do not have such an agreement, the person making the request must brew beer for the *umgqibelo*. In some cases the beer drink takes place immediately after the completion of the task; in others it takes place some weeks later. The quantity brewed varies from a medium sized pot (about fifty litres) to a full cask (200 litres).

Agricultural companies and their combination into *umgqibelo* are specialized work groups in that they involve the use of cattle for a limited number of tasks. There are two other, non-specialized work-parties, distinguishable from *isicelo* in that they involve generalized appeals for assistance rather than requests to individuals, and different from the labour-commodity exchanges discussed above in terms of their size and organisation as well as the reward offered.[11] Both types involve *amandwandwa*, the term used to describe the workers, and must be accompanied by maize beer. Other food is usually also provided. The smaller of the two is called *indwandwa yekhaya* ('a home group') or, more commonly, a 'congress' (*ikongresi*, lit: 'a gathering').[12] It is recruited from the local sub-ward sec-

---

11 In this paper I do not deal with the larger and rarer kinds of work groups, which I have been informed of but never witnessed. These include *isitshongo*, which is a combination of a number of ploughing teams (perhaps six or more) drawn from the whole of the sub-ward, and *ibhoxo*, a women's work group in which women from the entire sub-ward participate in collecting firewood, cutting thatching grass, etc. According to informants, brandy is normally provided at these events in addition to the beer and other food. In practice the term *isitshongo* is also sometimes used to refer to *indwandwa yelali*.

12 According to Kropf *indwandwa* means 'something valuable or scarce' (1915: 90). Other terms used to describe this group are *umtabata*, which Kropf (1915: 401) defines as 'a number of people who club together to work in a garden', and *igxabaxu*.

tion or from more than one section where such sections have an agreement to call on each other for work-party purposes. In the case of Ngingqi and Komkhulu sections of Folokhwe sub-ward, for example, people say that these two sections 'work together', and a Ngingqi man may ask part of Komkhulu as well as Ngingqi people to join his 'congress'. The background to this is that Ngingqi is relatively small and was once part of Komkhulu until it split off to form a separate section.

The larger type is called an *indwandwa yelali* (a sub-ward group) or simply *indwandwa* (Heron 1990). In this case all sections of the sub-ward are called on for help. The announcement that assistance will be required is made at some public gathering such as a beer drink. The size of these groups (about fifteen to twenty people for an *ikongresi*, thirty or more for an *indwandwa*) is also related to the amount of beer provided. The smaller event requires around sixty or seventy litres, the larger 120 litres or more. As with other beer drinks, word soon gets around how much beer is being prepared, and larger quantities attract more workers as well as more non-workers to the subsequent beer drink. These work parties may be called for tasks such as cutting thatching grass, hoeing, fencing, harvesting, and so on, but the commonest is for weeding. As with ploughing groups, although the pool from which workers are recruited may be defined in formal terms, participation in a work group is related to the actual relationships between people, based on close friendship, mutual interest, reciprocal obligation, and so on.

In discussions with people in Shixini, opinions varied on whether paying people to hoe was a more efficient way of getting the work done than organising a work party (*ikongresi* or *indwandwa*). For a work party the host has to brew beer and provide food for those who come to work, and this is costly, especially in terms of labour. Eight beakers of beer have to be made for *ikongresi*, and a cask-full for an *indwandwa*. In addition, there is no control over who will come along and who will not, since the invitation is a general one to the people of the local area. Astute householders, however, when holding a work party, utilise their kinship and other connections to try and ensure a good turnout, and of the right people. Some people complain that those who attend work parties do so primarily for the beer and do not work hard enough, but this is not a common complaint. Nowayinesi had organised *ikongresi* for the hoeing of her garden some time before she killed a 'pig for hoeing' for her smaller garden. She felt that the *ikongresi* had not worked quite as well, because the workers had not completed the task of hoeing the garden as she had hoped.

Another, older woman considered herself to be too old to brew and therefore, she said, if she did not have money to hire labour, she had to get people to help 'by means of some plans' *(ngamacebo)* such as killing a pig. But she added that 'I used to brew when I was strong enough' and that she thought *ikongresi* was a better way to get the work done because of the large number of people who would attend. Making 'plans' enabled her to get only a few people at a time, and the work thus took longer to be completed. Nobongile agreed that *ikongresi* was a better way of getting the work done, saying that those who participated worked longer hours than hired labour, and no individual could demand a particular amount of the reward. Providing an additional incentive for the workers, such as a bottle of brandy in addition to the beer, can help to ensure a satisfactory result. For example, Santikana held an *ikongresi* which was attended by 12 people, and for which he provided a full cask of beer as well as a bottle of brandy. The event was successful in that the entire garden was hoed and Santikana was pleased with the way the work was done. He said that it was because of this that the unusual step of providing brandy had been taken, as an extra reward given voluntarily, but it is likely that the workers knew that it was in the offing.

Heron (op cit) points out that exactly who organises and who attends work parties of this kind depend largely on the position of the homestead in the domestic cycle. Those who tend to call work parties are larger, wealthier homesteads, which are prepared to invest in food and beer for workers in the hope of a good return at harvest time. These are also usually homesteads with significant cattle holdings and a position in an agricultural company that enables them to cultivate their land at the optimum time. Those who attend are largely younger people and people without large gardens themselves, and who are not in a position to invest in work parties. Younger people attend work parties because they expect to have larger gardens and to call work parties in the future, and they wish to establish their reputations as good workers who are deserving of labour reciprocity at some future time. In the meantime, reciprocity from the wealthier households may be manifested in gifts or loans of maize or other items, including cash, as pointed out above, or in the form of cattle loans in terms of the practise of *ukunqoma*, whereby a person without livestock looks after some of the animals belonging to a wealthier person and receives one or two of the offspring as well as products such as meat and milk in return. This too, then, is a case where long-term imbalances of a kind are in evidence, and of short and long term reciprocity co-existing in the same labour institution.

The beer drink that follows the completion of the work, whether it is for *umgqibelo* or other types of work party, is like other, ordinary beer drinks in many respects, but to appreciate the significance of this a digression into beer drinking in general is required. We will then be in a position to return to the beer associated with work parties and to understand the link between co-operative work and beer drinking in general.

6

Drinking and working

The occasions for beer

What becomes evident from the above is that labour in Shixini, most of which is directed at agricultural tasks, is initiated by individual homesteads but dependent on assistance from others, almost invariably neighbours (who may include close kin) within the sub-ward or, more frequently, within the sub-ward section. This is explicitly conceptualised in terms of *ukuvana* ('friendliness' or 'mutual sympathy and understanding'), mutual helpfulness (*ukuncedana*) and good neighbourliness (*ubumelwane,* which literally means 'supporting each other'), and is based on an exchange in which labour reciprocity features prominently. People who live together in the same local area help each other and share resources with each other on a reciprocal basis, to the benefit of each individual household. To a large extent, beer drinking in Shixini is homologous with this pattern, and it has to be analysed in order for us to reach a fuller understanding of the homestead and its relationships with others in the labour process. Beer is brewed for a great variety of ritual and ceremonial purposes in Shixini, but only some of these events are 'beer-drinks' (*iintselo* or *iindywala),* in the sense of being events held in an individual homestead at which beer is the central focus and the only foodstuff available.

The events recorded in Shixini in terms of this definition include beer brewed for no apparent reason other than to give it away. This is euphemistically referred to by a descriptive term such as *isichenene* ('a little drop'), or *intselo* ('a drink'), though the amount of beer brewed can be large. There is frequently also a private and unannounced reason for holding an event of this kind, though often the motive is simply to provide hospitality, due to it being some time since the household last brewed. Such beer is thus also referred to as *utywala besintu,* (literally 'the beer of human-ness') a term which reflects its essentially moral quality in accordance with the meaning of *isintu*, 'human-ness'. This event therefore conveys a quality of moral righteousness, as the right and proper thing to do as a human being and member of a community, and we can immediately see the broad moral connection between beer brewing and co-operative work. Therefore, even if the reason for brewing is fairly private, it occurs within a context of the ideal of providing 'food', as the beer is called, for others. Unlike co-operative work, however, one cannot assume that people who attend the beer-drinking do so because they have a close relationship of some kind with the host; indeed, even people whom the host is not on good terms with may be there. No one is specifically invited, everyone in the sub-ward and in nearby sub-wards, and any passer by or visitor, is free to attend. The brewing process serves as notice that a beer drink is to be held, and people soon get to know when it will be ready for drinking.

Many beer drinks are brewed for a specific purpose and named accordingly. All beer-drinks are religious in the sense that the ancestors are thought to want their descendants to brew and to partake of beer in spirit, and some have a specifically religious motive. These include events held as a substitute for a ritual killing of an ox or goat, such as when beer is brewed to 'make known the homestead' (*ukwazisa 'mzi*). It signals the formal establishment of a new homestead, which comes into (social) being through the participation of neighbours and kin (including the dead) in the beer drinking in the home.[1] *Umsindleko* (a provision) is a beer-drink held to welcome a returned migrant worker and to thank his ancestors for their protection and blessings.[2]

Two beer-drinks are associated with death and mourning. The first is held for 'releasing the family' (*ukukhulula usapho*) from mourning, two or three months after the death of a senior family member, and to allow them to participate in community life again. If it is a male homestead head who has died, his widow remains in mourning for an additional month or two and is then 'released' from this status and incorporated back into society, in the new status of widow, again through the medium of a beer drink. In such cases the beer drink not only marks the transition but also constitutes it in the sense that a mark of having been released from mourning is the ability to mix freely with people and attendance at beer drinks. The same point applies to a beer drink held to promote a woman to senior status from her position as junior wife, which incorporates her into the ranks of senior women. Beer without a specific name may also involve distinct elements of status change. When *intwana nje* (literally 'a small thing') is brewed for a new homestead it marks the important new status of homestead head for the host and, in the case of a married man, a corresponding change in status for his wife.

There are a number of services for which it is appropriate to brew for a beer drink. When a homestead site is allocated, it is customary for the applicant (or his/her father/mother) to brew 'beer for the site', which serves formally to bring the allocation to the attention of the community at large and to signal their acceptance and approval. Similarly, a successful

---

1 Another beer-drink called *inkobe* ('boiled maize') is a similar substitute, held for a female ancestor, usually a deceased mother. This does not mean that boiled maize is provided for the guests in either the original ritual or substitute (i.e. beer) form. Rather, the word *inkobe* evokes the image of a woman and mother, as one who is associated with the provision of food. In addition, beer is regarded as food and often referred to as such. At work party beer drinks some of the beer allocated to workers is called *inkobe* (see below), and they are often also given conventional *inkobe* (maize) to eat.

2 Kropf (1915: 391) translates *umsindleko* as 'food which a woman prepares and keeps for her absent or traveling husband', and as a beer-drink it is always spoken of as being organised for the migrant by his father or, in the case of more senior men, his wife.

application for a new field is followed by 'beer for the field', after the formal marking out of the field's boundaries by local agricultural officers.

There are two types of beer-drinks at which most of the beer is sold rather than given away. The first, brewed by an individual homestead, is called *imbarha*, and is held to raise money for the homestead. The second is called *inkazathi*, a term commonly used in both town and countryside when people club together to buy something as a group. As a beer-drink it retains this common meaning, in that it is organised by a group of people who club together to buy the maize required for brewing, and to collectively undertake the preparation of beer for sale to raise money for communal purposes (e.g. the repair of school huts).

At both of these events, 'commoditisation' (i.e. monetization) of beer is restricted by a number of conventions reflecting, as Barth (1967, 156) noted in Darfur, a view of beer as 'a festive idiom of cooperation and companionship and not appropriate as an object of commercial bargaining.' Generally speaking, Shixini people do not approve of brewing maize beer for sale, and no self-respecting homestead head would like his/her homestead to be seen as a place where beer can be bought on a regular basis. This has to do with the status of maize beer as a highly prized commodity which ought to be made and distributed in a socially appropriate manner, and which always has religious associations. To brew for regular sale would be to dishonour the homestead, its occupants and the ancestral spirits associated with it. Brewing *imbarha* for sale in situations where a homestead 'has a problem' is acceptable, and allows for an exception to the general rule. Many households very seldom brew for sale, but a few (perhaps one or two percent) brew *imbarha* about once or even twice a year. These tend to be poor, female-headed households which engage in a variety of subsistence strategies, of which brewing for sale is one (see chapter two).

A major constraint on the sale of beer is that brewing requires a lot of labour and beer spoils rapidly, so it must be consumed within a day or two. In addition, *imbarha* beer, like any other, must be distributed according to a set of conventions enforced by the local community. A large portion of the beer has to be given out free of charge to those in attendance. In addition, there are rules that seek to ensure that all the territorial groups present (i.e. the sub-wards and local sub-ward sections) must be given an opportunity to purchase beer, so that no one will leave complaining that they got nothing to drink.

A variety of factors thus combine to restrict the sale of maize beer, and to prevent it from becoming a commodity that can be exchanged for cash, as it has become in many other parts of Africa. But the fact that its entry into the cash economy is limited does not mean that it has little 'economic' significance. On the contrary, it is central to the local economy in the sense that it is through the (free) social distribution of beer that relationships of economic interdependence are reflexively constructed, perpetuated and manipulated. It is to this distribution that we now turn.

Rules of beer drinking

Beer is consumed according to a strict code and an elaborate etiquette, which is associated with both distribution and consumption. Before a beer-drink commences the host apportions the beer amongst the various barrels and/or pots available, according to the various allocations that are going to be made. In doing so he/she consults with prominent male members of the sub-ward section, because the major portion of the beer will be distributed by the section and not by the individual host, as I discovered to my surprise when I first visited Shixini (see chapter one).

A certain amount is put aside as a reward for those who helped with the work and with the provision of utensils. No individual homestead owns all the barrels, pots and beakers needed for a beer-drink. As with the labour for brewing, its own resources have to be supplemented with items obtained from neighbours and nearby kin, adding to the perception of the drinking occasion as a community event. But the beer put one side for the workers does not go only to those individuals who helped; it is also distributed to close neighbours (members of the sub-ward section) in two non-public phases, one before the beer drink proper and one after the event itself has ended.[3]

---

[3] The first of these distributions is called *umlumiso* ('yeast'), the second, *isidudu* ('porridge').

The major portion of the beer is called the *intselo* (from *ukusela*, 'to drink'). This is divided into a portion for men and a (usually much smaller) portion for women. Separate portions are put aside for preliminary drinking, for gifts to specific people and groups, and for a variety of obligatory allocations to members of the local community. At an early point in the proceedings, individual beakers of beer are given out by the host to his/her close friends, certain kin (mainly affines), and to important people such as a visiting chief or headman. A beaker given in this way is called *umnono* (plural; *iminono*, meaning 'neatness' or 'carefulness') or *umahluko* ('difference' or 'distinction'). The people who are given this beer reflect actual networks of relationships, and vary across time as old links fade and new ones are forged. Sometimes two or three people are asked, rather apologetically, to share an *umnono* beaker, indicating that it is difficult to make choices.

A portion of the main beer is set aside for what are referred to as 'the customs' or 'privileges', amounting sometimes to as many as eight or ten beakers. This beer is also referred to as 'rights', 'duties' or 'expectations'.[4] The individual host gives out the preliminary beer and the gifts to individuals, but from the 'customs' onwards, decisions are made by the men of the individual host's sub-ward section and allocation is in the hands of the section's 'master of ceremonies'.[5] The 'customs' go to groups of people with whom the host section has a close neighbourly and geographical relationship — those who live in the same section of the sub-ward as the host does, and also those in other sections of the sub-ward. Such people are clearly seen as 'privileged', in that they are separated out from the others (those from other sub-wards) for the initial allocation of beer.

The 'customs' include a variety of separate, named beakers whose symbolic significance for co-operative work is immediately apparent. This symbolism is consciously acknowledged: People say that the 'customs' mark 'living together as a section' (*ukuhlala ngokwelali*), and that it indi-

---

4 The relevant Xhosa words are *amasiko* ('customs'), *amalungelo* ('rights' or 'privileges'), *iimfanelo* ('duties' or 'properties') or *iziqhelo* ('habits' or 'expectations').

5 The master of ceremonies is called the *injoli*, from the verb *ukujola,* 'to apportion or divide'.

cates 'the way we look after (lit: nourish) each other' (*indlela yokutyisana*). Some of these beakers are as follows:

i. *Iimvuko* (awakening or alerting) consists of one or more beakers allocated to men and women of the host's sub-ward section, and to the members of other neighbouring sections. The term is derived from *ukuvuka*, 'to wake up', and it symbolizes close neighbourly relationships. People who give each other this beer are those who call on one another for assistance, for example when there is trouble such as stock-theft at night. Sometimes the 'awakenings' beer is sub-divided within the section on the basis of sub-sections, and the allocation of beakers by one of these to the others dramatizes actual relationships of friendliness and co-operation or, in the case of choosing not to give to a specific sub-section, friction and hostility between them. In other, uncommon cases a section of one sub-ward may give 'awakenings' to a nearby section of another sub-ward, reflecting relationships of economic co-operation between the groups involved.

ii. *Isikhonkwane* (a 'peg' or a 'nail'), is a preliminary beaker whenever beer is sold, given free of charge to men of the local sub-ward as a whole, and is frequently referred to as the 'beaker of the spoor' (*ibhekile yomkhondo*) or simply as the 'spoor'. It is the people of the sub-ward, say informants, who will be asked for help if one has lost a goat or a cow. The beer secures ('pegs down') their assistance.[6]

iii. *Intluzelo* is also a distribution given out when beer is sold, if the amount brewed is large enough. Whenever beer is brewed a beaker or two are given to the women who did the work of grinding and brewing, as the brewing process is completed and the beer is being strained. It is called their *intluzelo* (from *ukuhluza*, 'to strain beer'). A distribution with the same name at *imbarha*, where the beer is sold, indicates that the host is grateful to those who have come to 'work' for the homestead by purchasing beer.

---

6 Wilson et al. (1952) recorded a beaker named *isikhonkwana* (*sic*) in Keiskammahoek district in the late 1940s. It was the first serving of beer at a work party for tasks such as weeding, given out shortly after the work had started, so as to 'peg people down to work'.

iii. *Ibhekile kasibonda* ('the sub-headman's beaker'). As the name implies, this beaker goes to the sub-headman in recognition of his status and the services he performs for people.

Great care is taken to ensure that these beakers are distributed as they ought to be, that is, in the right amounts and to the right people. Each section and sub-ward knows how many beakers it ought to get, and what size these should be, given its relationship with the host group, the pattern established at previous beer-drinks, the amount of beer available, the type of event, and the number of other groups present.

If the hosts are careless and make mistakes they are made to pay for it at subsequent beer-drinks, where their allocations will be less that they would normally get. Serious oversights or deliberate slights might lead to a fine being imposed on the offending group — usually in terms of beer that has to be brewed to make good the wrong, but if it is serious, or the fine is not 'paid', it can lead to a severance of the beer-exchange relationship between the two sections involved. This usually reflects some pre-existing tension between the sections arising in non-beer-drinking situations. The relationship between the distribution of the 'customs' and social practice is also indicated by the fact that adjoining sections of adjacent sub-wards who collaborate closely may give each other 'awakenings' to the exclusion of other sections of the sub-ward in which the beer-drink is held. Again, discrimination and choice are exercised, and the beer exchange as a dramatic performance 'indexes' everyday realities.[7]

There are two ways of giving out the main portion of the beer. One, *ukugabu,* is simply to allocate each beaker to a certain number of people, so that the distribution is fairly even. This method is uncommon and seems to be used only when there is little beer available. The other, preferred way of giving out beer is according to *ukulawula* ('to distribute by groups'). Before this can be done, a procedure known as 'the arranging of the hut' takes place, in which the host group allocates seating places (*ukunika iindawo*) to the various territorial, gender and age groups present. These places are inside the hut and, if it is full, also outside in the

7 On the 'indexical' nature of performance see Bauman and Briggs (1990) and Tambiah (1979).

courtyard. They are ranked in accordance with the conventional division and ranking of space in a Xhosa homestead,[8] and the beer is served to a group strictly at the place allocated to it. This is an extremely important process, since it is bound up with the status of each participating group and individual. Those with the higher-ranking places will be given their beer first, and will receive a larger quantity than other, lower-ranked groups. Again the basic principle is that local groups receive more beer than others, and before the others. Men get their beer before women, who occupy a lowly ranked place outside the hut, and seniors get theirs before juniors.

Certain fairly complex notions, crucial to understanding beer drinks, are embodied in the term *ukulawula*. This word means to 'rule', 'govern' or 'control', also to 'apportion' or 'divide', and to 'arrange' or 'give order to'. One of the contexts in which the term was used in the past was when a chief allocated parts of his chiefdom to his sons, as areas over which they would have jurisdiction. In this sense the chief maintained control and power over his sons, yet gave them shares of the chiefdom, and in so doing also provided an administrative and spatial order. *Ukulawula* at a beer-drink means that the hosts, as a territorial group, recognize their relationships with other similar groups, allocating both seating places and beer on this basis. Visiting groups have rights in this respect that they expect to have recognized. Rights are based on the reciprocal beer-exchange relationship between giving and receiving groups, which change over time as the overall relationship between territorial groups changes. In this sense the beaker allocated to any particular group (and also the beakers given to individuals) has a 'biography' (Kopytoff 1986). As in the case of other 'commodity paths', precedent is not always followed and diversions are frequent. Appadurai (1986, 17) notes that the flow of commodities is always 'a shifting compromise between socially regulated paths and competitively inspired diversions'.

The visiting territorial groups (that is, the other sub-wards and sections) at a beer-drink are referred to as 'oxen', 'spans [of oxen]' or, sub-

---

8 Kuckertz (1997) provides a description of the way space is divided and evaluated at a Mpondo beer drink; the Xhosa system is different and more elaborate (see McAllister 1987).

stituting a military for an agricultural metaphor, as 'regiments' or 'armies'. The host section is referred to as *mpi yakuthi* ('our people' or 'our regiment'), though there are many alternatives here, such as *amawethu* ('people of the home', a term also used to mean kin and ancestors in certain contexts), *iinkabi zalapha* ('local oxen'), or *iindlezana* ('most generous ones'). These terms reflect the great importance of neighbours in Shixini as people who ideally, and to a large extent in practice, work for and with each other (like oxen), who support and defend each other (like combatants), and who counsel and assist each other in various ways.

Despite the formal allocations, most beakers circulate widely. Any person, when it is his/her turn to drink, may call someone else, usually from another group, to have a few mouthfuls. There are no rules regarding who may or may not be called, and this practice allows those with beer to call men or women whom they may wish to acknowledge or to thank in some way. A person called for a drink may 'carry' another one or two people with him/her, to share in the invitation.[9] Because women are normally allocated far fewer beakers than men, they receive beer primarily through this informal sharing (*ukurhabulisa*), in a procedure that emphasizes their subordinate status. A woman called into the hut for a sip enters in a highly deferential manner, kneels some distance in front of the man who called her, stretches out to receive the beaker, and drinks her share, handing the beaker to the person she has 'carried' with her, before giving it back with appropriate expressions of gratitude.

In practice the informal sharing of beer means that after every formal allocation, in which great care is taken to divide the beer in terms of a particular structure, in terms of the meaning of *ukulawula,* and to allocate it to a particular place, the structure dissolves into a chaotic jumble of people coming and going between groups, entering and leaving the hut, accompanied by much shouting and merriment, as those with the beaker in their hands 'call' others to drink, and those called in turn call out to others to accompany them. It is an obvious manifestation of *communitas*.

---

9 This is referred to as *ukubeleka*, a term which in everyday speech means to carry a baby on one's back. Since babies are dependent on their mothers for food, the metaphor is appropriate.

The points that emerge from this description of the rules of beer drinking are clear. The allocation of places and beer and the way in which the beer passes from hand to hand, provides a map of the social order and the principles on which it is structured. Age and gender are relevant here, but it is territory that is most prominent, and within which age and gender are subsumed. Kinship, notably, does not feature outside of the gifts to individuals by the individual homestead head and host, where it is mainly affinal kin who are recognised with the gift of a beaker. The rules of beer drinking emphasize the solidarity of neighbourhood groups, especially of sub-ward section, but also of the sub-ward as a whole. The individual homestead has scope to acknowledge its individual relations with others, but for the most part it is embedded in the sub-ward section and has no existence outside of the section. Individuals are able to pass the beaker to others, as individuals, when it is their turn to drink, but they receive the beaker in the first place by virtue of being members of territorial groups. And those who are called for a sip bring others with them on the basis of *their* individual links, not the original caller's. Sections are portrayed as groups of people who assist, work for and feed each other, and which do so on a reciprocal basis in terms of the morality of neighbourliness and mutual helpfulness. This is not simply symbolized but also constituted through beer drinking, since it involves a system of reciprocal exchange and the consumption of the beer, a valuable commodity.

In this way, beer drinking as a performance creates information and new experience, it 'gives form to' diverse and disparate knowledge (Fabian 1990, 11-12). This knowledge is not readily accessible or expressible elsewhere, it is 'hidden' to people as well as to anthropologists. The drama of the beer drink is able to generalize, to bring together in a single form information from a variety of sources. Turner long ago pointed to the qualities of ritual symbols that allow them to do this, such as condensation and multi-vocality. One cannot see a territorial community like a sub-ward, or the complex web of relationships and interactions between members of such groups, but these can be constituted and visualized through the rules of beer drinking, the oratory, and the other symbolism of beer drinks. Beer drinking involves a dramatic spatial representation (and a constituting) of the community, the homestead, and of the nature of the co-operative relationships between homesteads as members of wider groups. As we shall see below, it performs this role in a specialised way at the beer drinks associated with work parties.

We are now in a position to return to co-operative work and to look at the beer drinks that are associated with work parties.

Work-party beer

Umgqibelo, as mentioned above, consists of a combination of agricultural companies called together for ploughing and planting. At the beer-drink following the work, the bulk of the beer is handed over to the senior men of the *umgqibelo* group, the *usonkabi*, literally 'the owner of the oxen' but the term refers to all the senior members of the company. It is these men who sit together in the senior place in the hut,[10] and who control the distribution. They have to decide how much to give to the people who did the physical labour, the young men and boys, and how much to keep for themselves and the other men and women (non-workers) who come along to the beer drink. They must show their appreciation to the workers and provide them with enough beer to enable them, in turn, to give beer to their contemporaries who come to the beer drink, but they need to retain enough to signal their status as *usonkabi* and to provide for the other senior men and women who will be present.

The beer is seen as being given to the senior men of the *umgqibelo* by those who laboured (their 'sons'), to distribute as they see fit. At an *umgqibelo* in Chibi section of Folokhwe, for example, the leader of one of the companies started off the proceedings by saying: 'Here is the point, people of the *umgqibelo* group. Your sons are back [from the fields] and they have given you this beer. You are the people who are going to distribute this beer; even the people [sitting] outside [i.e. non-Chibi men], it is you who will decide when to call them in.' He was making the point that the men of the company were in full charge of the event, and that in

---

[10] This is the right hand side of the hut (from the perspective of one looking out) against the wall next to the door. Where senior men are included in the group who laboured, they are grouped with the other seniors for allocation purposes and not with the labourers.

this they took on the responsibility normally held by the sub-ward section, namely that of regulating the relationship with other territorial groups through the allocation of sitting places and beer.

At these events the young men are normally given about half of the beer in a separate hut, and redistribute it among their peers as well as to the (uncircumcised) boys who helped them, who drink it in the same hut as the young men, together with other boys of the immediate neighbourhood. The normally strictly observed distinction between circumcised (men) and uncircumcised (boys) becomes secondary to common participation in a work group. Boys do not attend non-work party beer drinks, and only more senior young men, those who have been 'admitted' to beer drinking, do so. The *umgqibelo* beer drink proceeds in the main hut, and is attended by anybody who normally attends beer drinks and who cares to come along. The association between the working and the drinking is close: In the ploughing season, a person who says he is 'going to *umgqibelo*' may mean either that he is going to work as part of a company or that he is going to a beer drink associated with such work, or both. The basic distinction at this and other work party beer drinks is that between those responsible for the work and the rest, but the principle that people who labour should share the reward with others is clear.

At *umgqibelo* and other work-party beer drinks, those who were not involved in the labour are referred to as *abasarhi* ('beer-seekers')[11] or as *iingcanda* ('porcupines'). Like the latter, known for raiding gardens at night, they have come to 'eat' food (beer) which they have not worked for, echoing the morality of co-operative work itself — the workers will benefit not only the organiser, but also the others who come along for beer, though the allocation to the 'beer seekers' signals their contextually inferior status as non-workers. No matter how prominent their individual statuses may be in everyday life, here they must submit to the authority of the workers. Nevertheless, as in similar contexts in many parts of Af-

---

[11] From *ukusarha*, 'to go about seeking beer or brandy...' (Kropf, 1915: 381).

rica,[12] they have a right to be present and to be allocated beer, with the exception that a person renowned for not working for others may be asked to leave. Although those who do the work enjoy (temporary) high status and the major share of the beer — the two are inter-related — people from the local sections and sub-wards who have not participated are also given beer — they 'eat' by virtue of the work done by others. At a beer drink held for a group of young men who had cut trees for building, the beer was spoken of as being given to the seniors by those who laboured (their 'sons'), and one of the senior men suggested that these young men be thanked for their efforts. But a prominent local senior objected to this, saying: 'when a man works does he work for himself? Does he not work so that others can eat?'

The perception of people of the sub-ward section as a group who work together and support each other, which is the symbolic message carried very strongly at beer drinks of all kinds, is reinforced by the fact that no able-bodied person from a section in which a work party beer drink is held would normally show up at the event as one of the *abasarhi*. If they are to be present, they ought to be present as workers, not as 'beer-seekers' in their own section, and I have heard men lament the fact that they could not attend a work party beer drink because it was being held in their own section, and which they ought to have been present at as workers. This does not, of course, apply at *umgqibelo* beer drinks, where the membership of the working party is determined by membership of the agricultural companies involved rather than by section affiliation.

The senior men of the *umgqibelo* group also decide how much beer to give to the 'beer-seekers', both men and women. If there is sufficient beer, the 'beer-seekers' are divided according to section and sub-ward, and the 'customs' given out in the normal way, except of course that the host sub-ward section is formally replaced with the *umqgibelo* group. However, since *umgqibelo* is often made up of companies most of whose members belong to the same section, it is this section which, in effect,

---

12 See, for example, Donham (1985), Krige and Krige (1943), Kuper (1947). At Lovedu work parties, the non-workers were called 'solicitants' and they usually far outnumbered the workers. (Krige and Krige 1943, 54).

Photo 3
Members of work party drinking beer in the maize field in which they have been hoeing to remove weeds

controls the allocation, and which occupies the senior seating position inside the hut. Members of the host section who are not members of the companies involved should in theory occupy the next most senior position, but they may simply be asked to sit with the others. This is not always the case, and there are times when the *umgqibelo* companies are from other sections of the sub-ward or even from an adjoining sub-ward, though this is rare. In such cases, or when the beer is not a large quantity, the *abasarhi* are allocated a number of beakers, each beaker being for a certain number of people. This allocation only happens late in the proceedings, after the workers have had their share from which, of course, they call non-workers for sips.

The functions of the agricultural companies are important enough for Shixini people to ignore or reverse the principles on which status is determined in the need to acknowledge the role that they play. The blurring of the distinction between circumcised and uncircumcised, a basic one in many contexts of life, has already been noted above. In another case recorded, a relatively junior man, strictly speaking still an *umfana* (a young man), was given the responsibility of allocating the beer because he was the leader of the company involved in ploughing and planting the homestead's land. This involved the homestead of Nontokayise, an *inkazana* (a woman who has never married) who lived in Ngingqi section of Folokhwe with her two married sons, their wives and children, and two unmarried daughters. Her sons were usually away at work and her father, Sihota, who lived nearby, played an important role in the affairs of her homestead. In 1983, at Sihota's request, her garden was ploughed by a Komkhulu ploughing company led by the young man, Ncedabantu, who had recently inherited the homestead and cattle of his father, Ndoda, who had been the leader of this ploughing company until his tragic death in a mining accident. The oxen used were those of Ncedabantu, Stokwana and Gulakulinywa, all of different clans and lineage segments, but all resident in the Komkhulu section of Folokhwe. The beer for the company was brewed some weeks after the work had been completed. Since neither Sihota nor his daughter were members of this company, beer had to be brewed for it. As Mzilikazi put it, 'it is 'soup' for the people who ploughed'. The proceedings started only once Ncedabantu had arrived and seated himself at the back on the right hand side, the appropriate place for a young man who had only recently started attending beer drinks. Sihota, representing the homestead, stood and addressed Ncedabantu, saying that the beer had been brewed for those who had ploughed. Preliminary and *amasiko* beakers were drawn and handed out under Sihota's direction. After the *amasiko* had been consumed, Molusweni (the regular *injoli*, or 'master of ceremonies' for Ngingqi section) stood to say that he had been instructed by Sihota to draw a beaker for Ncedabantu. The beaker was placed in front of Ncedabantu, and circulated. Molusweni also said that the beer had been brewed for those who had ploughed the garden, and indicated the amount (a pot full) that was to be handed over to the workers. The beer was then formally handed over to Ncedabantu.

Ncedabantu gave a short reply and started circulating the beaker drawn on Sihota's instruction. He then consulted Gulakulinywa and Molusweni about how to divide the beer. Stokwana and Qamela (the Komkhulu injoli) also joined in this discussion. As the *usonkabi* (the owners of the oxen), Ncedabantu and Stokwana (Gulakulinywa was away at work) had to decide how to allocate the beer to others. Those from Chibi section who were present had no say in the matter at all, and waited patiently for a decision to be made. Stokwana then announced that they were to go ahead with the distribution. The beakers were drawn and placed in front of the door, and Qamela announced how they were to be distributed. Later a separate beaker was drawn for the young men (*abafana*) who had helped with the work.

Beer for other kinds of work party is handled according to the nature of the work done, with the major allocation going to the workers. Certain beakers are given out in the course of the work itself, by the homestead head. With a 'congress' for hoeing, for example, early arrivals are rewarded with a couple of beakers soon after they start working. These are called *emvikweni* ('on the verge'), after *umviko*, the strip of uncultivated land between different fields, to which the workers retire for rest and refreshment every now and then. There are a variety of other, named beakers, which we need not elaborate on,[13] except to mention that even at this early stage there is often a beaker for onlookers, called the 'porcupine' beaker.

Once the work has been completed, the workers retire to the host's homestead, where they are given beakers called *inkobe* (lit. 'boiled maize'), which alludes to the fact that workers are usually also given food other than beer, often including boiled maize. The bulk of the beer is then handed over to the workers, who divide it among themselves, usually according to age and gender (if both old and young, or men and women worked) and/or the numbers present. There are obvious parallels here with ordinary beer-drinks, where it is the section that controls the beer, and its distribution symbolizes, among other things, the general nature of interdependence between territorial groups. However, the immediate context is superimposed on the general rules, and the nature of the control and distribution is linked to the specific work done and the workers involved.

Here too, everyday status relationships may be reversed. At ordinary beer drinks it is the sub-ward section, the group that ideally and to a large extent in practice works together, that is in charge. If a beer-drink is held to reward a group of young men for a task performed, it is they who are in charge of the beer distribution, allocating a portion for the elders to drink. If women form a work party to cut thatching grass, they occupy the senior seating positions (sometimes making men sit outside, which is what the women often have to do at beer drinks), allocate a limited amount of beer to the men, and 'call' the men in for sips every now and then, amid much teasing and laughter at this reversal of the norm. The normally strictly subordinate and inferior status of women is thus at least temporarily suspended.

In June 1986 Ndabankulu (Chibi section of Folokhwe) was in the process of building a new hut, and his wife was responsible for providing the *idobo* grass (*Cympopogum Validus*) for thatching. To this end, she asked her neighbour and friend, Notatile, to organize a work party of

[13] Later beakers may be termed *umnyenyetho*, defined by Kropf (1915: 306) as beer 'given to workmen', after *ukunyenyetha*, 'to give a small quantity of beer to one's workers, shortly after they begin working'.

women in order to meet on a certain day to cut the grass required. Twenty women, all from Chibi section, agreed to participate on the date agreed upon. On the day the women met at the appointed place and cut the grass required. In the course of the morning seven beakers of beer (which had been brewed by Ndabankulu's wife) were sent down to the workers. The first two beakers were called *emvikweni*, for the early arrivals (see above) and the second two, a little while later, were designated as *umnyenyetho* — defined by Kropf (1915, 306) as beer 'given to workmen', after *ukunyenyetha*, 'to give a small quantity of beer to one's workers, shortly after they begin working' (*ibid*).

When the women returned, having completed the work, they went into the hut in order to decide how to divide the beer, all of which was regarded as being theirs. As Ndabankulu put it, *silawula ngabo,* 'we [men] are controlled by them', and the ambiguity of the term *lawula* within the given context allows one to translate this as both 'given beer' and 'controlled' or 'ruled over'. Having decided what to do, the women called the female 'beer-seekers' inside from where they had been sitting next to the woodpile, but the men remained outside next to the cattle byre, there being no space for them in the hut. A beaker designated as being for the sub-headman was given out and used by men and women, and men were allocated two beakers of *usingantselo* ('this is not a beer drink') which, like the beaker for the sub-headman, was part of the *amasiko* ('customs') beer. Its appellation indicated that they should not expect too much! Later, men were allocated a further two beakers, but the bulk of the beer was kept in the hut and given out to women, on the basis of numbers rather than territorial groups. After every allocation of beer men were called in for sips, amid laughter and sarcasm from others at this reversal of the usual practice. Only where a younger, junior woman wished to give a sip to a considerably older man was the beaker taken out to the latter by the former. Later in the evening, when it became cold and some of the women had left, men were permitted to sit inside the hut.

When men and women work together as *amandwandwa* (e.g. at hoeing parties), both are regarded as 'owners' of the beer, and although the decisions about distribution are made by the men, they have to be careful to ensure that the women are happy with the way it is done. When there is sufficient beer, each gives to the *abasarhi* of their own sex, and they make this allocation partly on the basis of how much they received in the past, and how much they would like to receive in the future, when the position of workers and non-workers was/is reversed. On the occasion recorded below male and female 'beer seekers' were treated as a single group, and the women workers were not happy with this, deciding to give half of their share to the non-working women.

In January 1986 Ngangehashe (Ngingqi section) called a 'congress' in order to have his field hoed. There had been a good bit of rain, and the growth of weeds was choking the maize plants, which were starting to turn from green to yellow as a result. He brewed two pots of beer (about 20 beakers or 150 litres) and put out the word to Ngingqi people, as well as to some Komkhulu households, that workers were required on a particular day. On the appointed day, fifteen people turned up to help Ngangehashe and his wife — six women, all from Ngingqi, and nine men, three from Komkhulu and six from Ngingqi.

At 7 a.m. one beaker was taken from Ngangehashe's homestead to the field for those who had arrived early. After clearing the first *isiqith*i (strip or patch) the workers retired to the grass verge (*umviko*) between Ngangehashe's field and the next to drink this beaker, which is called the *emvikweni* (lit: 'on the verge') beaker. This beaker is meant to serve as an incentive for people to arrive early, though those who arrived while it was being consumed were also given sips.

In the course of the morning a further four beakers were taken down to the workers, who were finding it tough going because of the dampness of the soil. Towards midday, men and women of Komkhulu and Chibi sections started to arrive at the homestead, in the hope of getting some beer. There were no *abasarhi* from Ngingqi, as it would have been unacceptable for a Ngingqi person not to join the work party as requested but to arrive for beer nevertheless.

At the homestead, one beaker for the *abasarhi* and the sub-headman's beaker were given out. These beakers, plus the *emvikweni* beaker, and two which were called *inkobe* (lit: 'boiled maize') given to the workers at the end of the day, were said by Ngangehashe to be the *amasiko* ('customs'), distributed by him, as homestead head. Once these two beakers for beer-seekers were empty the latter went down to the field, having been told that the rest of the beer would be allocated there.

Three of the more senior men, including Ngangehashe's father's brother, remained behind. At 1.45 p.m. two plastic containers containing five beakers of beer (a total of 35 litres) were taken down to the field. The workers laid down their hoes and sat down on the verge, and the beer-seekers came over from where they had been waiting. The spatial arrangement of the various groups emphasized the division between workers and non-workers, though there were divisions within these groups based on sex and age. Because the workers had included young men, they too were allowed to be present, but formed a separate group. An argument developed among the workers concerning a certain woman, Nophakamisa, who was sitting between the men's and women's groups. Qamela (the most senior male worker) told her to go and sit with the women, but she refused, and answered him indignantly. Other men started belittling her, saying 'Is this Nophakamisa a man?' 'We will give you an overall to wear', 'Are you still answering back?' 'Bring an overall — we are going to give you a pair of trousers', and so on. A number of men told her to move, but she responded that Qamela had no right to tell her to move, because he had arrived late. She pointed to a young man sitting in a similar position, near the women, and told Qamela to be quiet. Eventually one of the senior Ngingqi women called Nophakamisa over to the women's group and suggested that she sit down there. The fact that Nophakamisa dared to argue with a senior man at a beer drink, and got away with it, seemed to indicate the relatively equal status of men and women at the event.

A young man rinsed out the beakers and placed them next to the beer. Dyakalashe (Ngingqi section) and Qamela (Komkhulu section), the two senior male workers, discussed the question of distribution, and Dyakalashe stood to speak to the beer-seekers, addressing himself to the senior Komkhulu male among them:

Dyakalashe: 'Oh, Nojikile! Here is a word, which I want to put to all of you. The reason we are here is because we have been working, with these ten beakers. We have used five of them already. Now there are these here, and they too are five. Here they are, we are going to use them. Allow me to stop at that point.'

Nojikile: 'Oh, we are grateful for your words; that is the custom [*ngumthetho lowo*].'

Dyakalashe: 'Eeee, the important point is this. As porcupines, you would get your own porcupine beaker [*ibhekile yakho yobuncanda*], an *iqhwina* [a 10-litre beaker], if this was a full cask. That is not available now because this is a small affair. We will give you only an *inxithi* [a 7-litre beaker].'

Nojikile: 'No, fellow, it is understandable (*kuyavakala*); this is something attended to by the workers.'

One beaker was then given to the 'porcupines', one to senior male workers, and one to female workers. The latter were eating a dishful of boiled maize provided by Ngangehashe, and did not touch their beaker. After some time, and having discussed it among themselves, the senior women said to Dyakalashe and Qamela that they were not happy about the beaker for the porcupines. They had noticed, they said, that it was being used by the men only, and wanted to know if it had been announced if it was for both men and women, as it should have been. Qamela replied that it was not necessary to make such an announcement, as it was not usually done. The women, however, insisted that this was a mistake. Now, they pointed out, the porcupine women had no beer, and the result of this would be that when those women next controlled work party beer, they would not share it with the Ngingqi women. Dyakalashe offered to apologise to the porcupines for this, but the Ngingqi women asked him to give half their beaker to the other women instead. This was done.

When these beakers had been consumed, with the beer-seekers frequently being called for sips by the workers, another two beakers were distributed, one each to male and female workers. This time the men's beaker started with the young men, from where it moved back to the seniors. Again, beer-seekers were frequently called for sips. When the beer was finished, Ngangehashe persuaded the workers to complete one more patch of weeding, and the others drifted off home. The workers did not complete the entire field because of the dampness of the soil, but before they left they went up to Ngangehashe's homestead, where they were given the two beakers of *inkobe* ('boiled maize') beer — one for men and one for women.

This case illustrates a number of points relevant to a consideration of beer drinks as social dramas or 'performances'. The beer drink took place at two points — homestead and field — a spatial dichotomy that served to dramatise the dependence of the homestead and its inhabitants on the labour power of others, recognized at both points in the allocation and control of beer. Furthermore, the beer drink took place in accordance mainly with the fact that a job of work had been done, and that the relevant categories of people were the workers and the non-workers, rather than territorial, sex and age divisions. Although the latter were still in evidence, they were played down in comparison with other types of beer drinks. Young men who did not normally attend beer drinks were permitted to participate, and received beer to drink as part of the men's group. Women received a relatively large proportion of the beer, though still less than the men, whom they outnumbered. The normally strictly subordinate and inferior status of women was thus at least partially suspended, as illustrated also by the fact that Nophakamisa dared to argue back when told to move, the women's questioning of the beer division, Dyakalashe's admission that he had made a mistake with the announcement regarding the beaker for beer-seekers, and the changing of the allocation at the women's request.

The important point about work party beer drinks, then, is that they are modelled on the work done and are also a manifestation and dramatization of a wider moral principle, i.e. that members of the community ought to work together and help each other. The rules of beer drinking are modified to recognize the particular workers involved, while at the same time invoking the more general notions of reciprocity that are in evidence an any beer drink. So the work party as such can be seen as just one particular aspect of intra-community relationships (with 'community' being defined relative to the occasion, but seldom extending beyond the level of sub-ward). They do not contradict the picture of social order created at ordinary beer drinks, but invoke and highlight a particular aspect of it. The inclusion of *abasarhi* (beer-seekers, porcupines) in the proceedings makes sense in terms of the wider moral principle and the general interdependence between homesteads in the sub-ward. They are given beer because they too might be called upon as workers in the future; because they too, as homestead heads or wives of homestead heads, will brew for work parties. At ordinary beer-drinks the balance tips the other way, with the *general* principle of mutual helpfulness being emphasized, and the actual relationships in individual homesteads are involved being secondary, though still recognised in an abstract way through distributions such as the beakers called *iminono* (see above).

The rules of beer drinking thus define the homestead as a moral entity in both types of beer drink. Kuckertz (1997) shows that Mpondo beer drinking and the rules associated with it are conducted in terms of the idiom of 'respect' (*ukuhlonipha*), and provides an opportunity for modelling morally appropriate behaviour and the nature of social relationships. The rules governing beer drinking are instituted to ensure that there is proper 'respect' between the parties involved, but this 'respect' is not in itself constituted by these rules, which are simply a model of how behaviour ought to take place. '*Ukuhlonipha* is an *idiom* and as such a *model* for moral reasoning about human relationships.' (*ibid.*, 337, emphasis in original). It is an idiom 'used to think about the content and form of human relations in an ultimate sense' (*ibid.*, 340). The elements of which the rules of seating consist — neighbourhood, gender, and seniority — are the elements that allow people to structure their thinking about moral norms. My own view is that beer drinking is something more than this, it is itself the embodiment of important moral principles.

Whether a beer drink is held for a work party or not, however, the principles on which it is structured remain the same. This is clear from 'beer for harvest'. The agricultural work done by work parties, or by individual agricultural companies, has a sequel some six months later when the beer drink to give thanks for the harvest is held. Although not a work party beer drink, 'beer for harvest' (*utywala bo'vuna*) is connected with agricultural work done. The main purpose of this beer drink is to give thanks to the ancestors for the harvest received (although this is not always very explicit) and to ensure that the following harvest will be as good or better. If the homestead has only a garden and no field, a somewhat smaller affair may be held, called simply *isichenene* ('a small drop') at which it is explained that the beer has been brewed in order to give thanks for the harvest. In yet other cases, where the homestead has oxen that worked in the fields, independently of others or (as is more usual) as part of a ploughing company, this event is sometimes called *utywala beenkabi* ('beer for the oxen').[14] Nowayilesi held an *isichenene* in July 1982, using some of the maize that she harvested from her garden to brew in order to give thanks for the harvest. This is what was explained to the people who attended, but Nowayilesi said to us privately afterwards that one 'thinks for oneself' before brewing, and that this beer had been brewed to 'invoke the ancestors' of the home (*ukunqulela abantu abadala*) so that the following harvest would be even better. This beer, she said, would make her children strong and build up the homestead, by ensuring a good crop next year. Another woman said that such a beer drink was held to praise (*ukubonga*) the garden and to call (*ukubiza*) maize for the following year.

A similar event takes place among Mpondo, called *ukuselela iinkabi* ('to drink to the oxen'). It is held for the owner of the oxen (from another homestead) who ploughed and reaped, and he is publicly praised for his assistance at the beer drink (Kuckertz 1985, 201-202). In Shixini the emphasis is sometimes on the oxen rather than on the owners. Kabilawo said: 'you are praising your oxen if you had a good harvest. They have made it possible for you to eat well; you received maize. We say we are soothing the weals, as we often beat it as it was ploughing...praising the

---

14 Hammond-Tooke (1962, 144) recorded a similar practice among the Bhaca, called *ijiki leenkabi* ('the beer of the oxen') and Hunter (1936) recorded almost identical practices among the Pondo. After harvest each of the homesteads in the ploughing 'combination' brewed beer, the 'beer of the oxen'. 'The general public attend, but those who supplied equipment for the combination have the privilege of giving out (*ukulawula*) the beer' (1936, 88-89).

beast for its work has been the custom since we were born. You praise your beloved ox (*uyibongoz' inkab' akh' oyithandayo*).' Stokwana referred to this beer drink as *isabokhwe* (a 'shambok' or rawhide whip): 'The oxen were beaten and it was painful. Now we should praise it because it has worked.' When Ntanyongo brewed for his oxen, Ngcinibandla said to the people that 'this beer is for the ox called Komdele [the lead ox] ... he is praising this ox for having ploughed and planted, and for taking the maize from the fields.'

In view of the close association between cattle and the ancestors among Xhosa people, and the fact that ancestors are sometimes referred to as 'cattle' in invocations to them (McAllister 1979, 121-122), it is likely that 'oxen' here serves partly as a euphemism for the ancestors. In 'praising the oxen' (*ukubongoza iinkabi*) or 'praising the field and the hooves of my father's oxen' (*ukubonga idiza namangina eenkabi zikayise*) Shixini people are also thanking the ancestors. Some say that if this beer drink is not held, 'there will be no harvest next year', although others say that 'nothing bad' would happen.

Nontaka's wife decided to brew beer for harvest in November 1976 while her husband was at work. The other members of the Tshezi agnatic cluster (Nontaka's brothers) were also away, so the beer drink was run by a neighbour and matrilateral kinsman, Ngcinibandla, of the Cirha clan, who is also the head of the ploughing company that Nontaka belongs to. The event took place at Nontaka's homestead in the Komkhuku section of Folokhwe sub-ward. Nontaka had received at least eight bags of maize from his garden in June, and Ngcinibandla said that the brewing was done 'so that there should be ten bags next year.'

Ngcinibandla explained the purpose of the event to the men shortly after they had arrived and seated themselves next to the cattle byre:

'Here is the reason why I have asked for your attention, my people. Here at home brewing has taken place, beer for harvest. It is brewed by Nontaka, in praise of the oxen which carried and ploughed and so forth. So, these are the beakers for the harvest, these three. I stop there.'

These three beakers were given to the people who did the ploughing — other members of Nontaka's ploughing company, nearly all of whom were Komkhulu men, and to the women who did the reaping, in recognition of their efforts on behalf of the homestead. The men received two of the beakers, the women one.

An argument developed about where the beer drink should be held. Ngcinibandla wanted to hold it outside, since there was only one hut, but other Komkhulu men overruled him, saying that it should take place inside. After some time the men entered the hut, which had now been made ready, and a further explanation was given by Ndoda, Ngcinibandla's older brother:

'Chiefs, we have arrived at the place that we were quarrelling about [the hut], there is no problem after all. There is the beer, the beer for harvest. That cask, my brother, you say it is for women? [Ngcinibandla confirmed that it was]. This full cask is the women's, this one is ours and this one is the women's. This one is exclusively for us [men]; he says then chiefs, he is giving it out to you Folokhwe people. I stop there now.'

Nduna then spoke on behalf of Chibi and Ngingqi:

'Let me say, chiefs, son of Mangono [Ndoda], it is a good thing for you to explain it here with your own mouth. You have announced it, as you are explaining it now. This man says I am complaining about him, even though I am he. Chibi and Ngingqi here are grateful for that speech. You are right; you are the brewer. [Referring to Ngingqi] People of Ngingqi, this is what it is about. This is the home of Sobashe [Nontaka's paternal grandfather]. All right, I am going to get started now. I do not want it to be said that the oxen were drinking just anything. No, I am drinking here at Sobashe's place. We are thankful for that now. Son of Sobashe, you have done something like this, because you promised those oxen in your company that you would brew for them. That is a fine thing; the whole of Folokhwe here says so. We too say 'Camagu!' to that speech of yours, and we are grateful. Chiefs, there is nothing else to be said unless it comes from Komkhulu, Chibi or Ngingqi. You have explained it and we can see it, man, the women's [cask], and the men's, we see it. Could there be any other man who might answer like this?'

The religious nature of the occasion is clear in Nduna's speech, particularly in references to Nontaka's late father's father, Sobashe, the association between oxen, Sobashe, and the homestead, and the use of the term 'Camagu', meaning 'be appeased', a phrase associated with religious occasions. Sometimes the religious element at 'beer for harvest' is more explicit than in the above case. When Zimene brewed 'beer for harvest' he expressed it as beer 'for the garden'. Sihota spoke and said that Zimene had brewed 'to praise the garden, so that tomorrow he will pluck a maize cob from this garden which is being praised. We say so too, we are praising. It is not us who are thankful, it is those who are not here!'. Other speakers at this event were also explicit, referring to the fact that Zimene had said that he had brewed 'for the children who ploughed', but that he was in fact doing it for his ancestors, and that he was invoking (*ukunqula*) them (see McAllister 1981, 12).

At 'beer for harvest' those responsible for ploughing the homestead's garden and/or field are given special recognition. The senior male members of the company that did the ploughing sit with the host section. Thus they share the 'awakenings' and other beer that is allocated to the section, and are also given a number of beakers in recognition of their contribution. The young men who worked, too, are given a share, which they drink in a separate hut. If no beer was brewed earlier for the people who helped with the harvest itself, they too are given a special allocation of beer when beer for harvest is brewed. A good proportion of the beer may be disposed of in this way, the rest being given out to sections and sub-wards in the normal manner.

As far as the relation between beer drinks and social practice is concerned, the important point about beer for harvest in Shixini is that it symbolizes the actual networks of cooperation and mutual help on which the homestead depends for a successful harvest, gives recognition to the individuals involved, and provides beer for the community at large, which thus receives 'food' by virtue of the host homestead (especially when the homestead itself laboured and provided the oxen) and the other people who worked. Yet it is not a work-party beer drink in that it is not tied directly to a particular episode of work. The workers who are honoured at it may have been involved in a variety of tasks, and not all of them would have been involved in the same tasks. In addition, 'beer for harvest' is brewed whether or not a ploughing company was responsible for some of this work. It is also important to note also that the status of *abasarhi* does not apply at beer for harvest. Those who attend do so as people of the sub-ward, or other sub-wards, not as 'beer-seekers'.

What is especially evident at beer for harvest is that beer drinking envelops or incorporates various forms of co-operative work. The ploughing group and the group of harvesters get a special share of the beer, in the same context where the same people, or some of them, also receive beer as members of the section and sub-ward, as close neighbours (in the initial and concluding phases outside of the beer drink proper), possibly as particular friends or affines (in the *iminono* distribution by the individual host), and certainly as individuals in the informal sharing that takes place after each formal distribution. Within a group of sub-ward section women, for example, allocated a beaker of beer as part of the women's share, one finds a dense network of relationships based on work (as well as on other things), which link them to each other, though not exclusively so — they are also linked to others across section and gender boundaries. In such a group are women who participate in each other's *isicelo* work groups, who engage in labour-commodity exchanges with each other, who belong to the same *umbutho,* who join each others 'congress' and *indwandwa,* and so on, and who brew beer for each other from time to time. Conceptualised in terms of social networks, the participants at a beer drink are engaged in multiple and overlapping sets based on exchanges of labour and beer, as well as on a multitude of other social and economic factors. Small wonder, then, that the emphasis in beer drinking is on generosity and harmony, that violence seldom threatens and is dealt with speedily and effectively when it does, and that the frictions and conflicts of everyday life are glossed over.

7

Beer, labour and the homestead

One way of analysing the relationship as well as the parallels between beer-drinks and co-operative labour is through the individual homestead, returning to the general theme of 'building the homestead'. Following Englund's (1999) example, it is helpful to conceptualise the homestead as a composite entity. It is evident that both beer drinks and co-operative work constitute the homestead in a particular way. As noted above, the basic unit of production in rural Transkei is the household, but its productive activities, as has been clear in the previous two chapters, are not simply a material matter but also a social and moral one. The individual homestead as the productive unit is inseparable from the social relationships that constitute it. In the course of agricultural and other work, and as a participant in a beer-drinking circuit, it draws on and contributes to other homesteads with which it has relationships, and in so doing constitutes itself as a social entity, while its inhabitants are constituted as moral persons at the same time. This explains perfectly the otherwise puzzling fact that in Shixini, even people who do not drink beer both attend and host beer drinks, call people for sips when it is their turn with the beaker, and are called by others in return.

The general moral precepts in terms of which the homestead's relationships are constituted are evident at certain stages of beer drinks; the actual relationships on which collaboration is based become evident at other stages, or at the beer drinks linked directly to co-operative work. But even at the latter, the actual relationships are presented within a wider moral context in terms of which homesteads exist and depend. The morality involved is encapsulated in a Xhosa term which was introduced in chapter one and which occurs in a variety of contexts, *ukwakh' umzi*, meaning 'to build the homestead'. As noted there, if one asks people why they hold beer-drinks, they reply that it is to 'build' the homestead. Migrant workers who spend lonely months working away from home in the mines or industries of urban South Africa, are praised as people who have gone to work to 'build' their homesteads, or their fathers' homesteads. At the ritual offerings of oxen and goats to the ancestors by lineage groups, people speak of how this will 'build' the homestead by securing the ancestors' goodwill and favour.

As these three examples indicate, this ideology of 'building' the homestead has social, economic and religious connotations. What soon becomes apparent if one pursues these ideas about 'building' is that these three components are always interrelated. Going out to work as a migrant, and successfully practising agriculture at home, provides the homestead with material assets that enable it to enter into a mutually satisfying relationship with the ancestors, through ritual, and with the wider society. Performing a ritual for the ancestors will bring forth blessings that are manifested in material and social terms, and the ritual itself involves a series of social, economic and religious actions. Using the homestead for the benefit of others will elicit reciprocal benefits and political support, and is pleasing to the ancestors not only because it is morally good, but also because it ensures the economic well-being of the homestead, and thus that future ritual obligations will be met.

The existence of the homestead as a social entity is evident in many contexts. It is as a result of the consent of the community that a person obtains the land necessary to construct a homestead and its garden, and it is as a result of brewing beer for the community that a homestead is socially recognized as an independent unit shortly after its establishment. Through this beer drink, the homestead becomes 'known' to the community and established as a social entity, as well as becoming 'known' to the ancestors through the participation of the community. It is pleasing to the ancestors for their descendants to 'build' the homestead in the economic sense, and it is clear that this would not be possible without relationships with people. As far as the ancestors are concerned, the very presence of people in the homestead is good, the 'noise' they make indicates that the homestead exists as a social entity in its relations with others, and it is this that makes beer drinks religious — the presence of people talking, arguing, smoking and spitting on the floor. The 'noise' of sociable beer drinking that was alluded to in chapter one is referred to as *ukunxola*, which in one sense means just that, a 'noise' or a 'commotion'. But it is also a term used to refer to an invocation to the ancestors in some contexts.[1] The commotion alerts the ancestors to the fact that there is something going on in their homestead, and thus ensures their attention and their presence in spirit. Similarly with smoking, supreme symbol of relaxed commensality, and with spitting; the more smoking, spitting and argument that there is at a beer drink, the better. It is in this way that the ancestors know that people are present in the homestead.

As it progresses though its various stages of development the homestead again invokes the participation of people from other homesteads to mark its transition from one kind of entity into another. The death of a homestead head, for example, requires the participation of the community not only in his/her burial, but also for the successful transition of the homestead into another phase, when the living are released from the restrictions of mourning and incorporated back into full social life. This is most marked at the beer drink to release a widow from mourning, and to redefine the nature of her homestead and its relationship with the community.[2] It applies also at other beer drinks marking transitions, such as the

---

1 For example, a man in distress may approach the byre gate at any time and invoke the shades, asking for their attention and help. This is usually known as *ukunqula* ('to invoke') but also as *ukunxola*.

2 For a detailed analysis of this beer drink see McAllister (1986).

promotion of a woman to senior status. In both cases it is the community into which the woman will be re-incorporated which participates in the event, and which effects the change by doing so.

It is through people as representatives of other homesteads that the potentially disruptive and threatening labour of a migrant worker is socially approved, and the benefits of migrant labour harnessed and made meaningful, at the *umsindleko* beer drink held to celebrate the migrant's return and re-incorporate him into the community. At this beer drink, in particular, it becomes apparent that the migrant's time at work, to be meaningful and moral, have to be directed at 'building' the homestead. But it is also clear that it is others, through their participation, that enable this aim to be achieved. They have an *interest* in the 'correct' interpretation of migrant work, and in the contribution that the homestead being 'built' makes to the wider community (McAllister 1981).

At all of these events, as indicated by the above details of how a beer drink proceeds, it is the homestead's relationship with its neighbours in the sub-ward section and, to a somewhat lesser but nevertheless important extent, the sub-ward, which is defined and acted out, and in terms of which its status as a homestead is constituted. People are implicated in the 'building' of every homestead,[3] which is defined in terms of its position in space and constituted in terms of its relations with others. It is in terms of this definition and the morality of mutual helpfulness that it receives the help and support of others, and its reciprocal obligations carried out. Returning to beer drinks in general, whether brewed for a work party or not, it is evident that they are a major manifestation of the obligation of every head to 'build' his/her homestead, and that it is at beer drinks that the three dimensions of 'building', the economic, the social and the religious, come together in a mutually-reinforcing manner.

Looked at in the broader context of homestead production, however, two seemingly contradictory tendencies are apparent. On the one hand,

---

[3] Among the Iteso (Karp 1980, 88-89), the very term for a neighbourhood group is derived from the verb 'to build', implying that this is a group of people who aid and work for each other in many ways. The same applies to Zulu speakers, where 'neighbours' is *abakhelwana* , literally 'those who build for each other' (Doke and Vilakazi 1972, 4-5).

co-operative and other labour as seen by Shixini people is based on moral norms such as kinship solidarity, mutual helpfulness, good neighbourliness, human-ness, and the like. In accordance with these, kin and neighbours collaborate extensively in a variety of ways, destitute households are provided with help without expectation of material return, and long-term imbalances are tolerated. At the same time, different segments of a kin group may work separately, a member of a ploughing company who does not pull his weight may be expelled, not all neighbours are asked to join an *isicelo* work party, and a person who refuses to join work parties may be ostracized. What beer drinks do is to juxtapose these two realities.[4] Drinking allows people to construct the world as it is, and also to construct an ideal world. This does not mean it is a false world, though it is a fragile one (Douglas 1987, 11-12). To express this somewhat differently, beer drinking contrasts the moral and the material and enables interpretation of the material in terms of the moral, while continually presenting the possibilities for the material contained in it. It dramatizes the groups and categories on which social life is based, and the moral norms associated with this, as well as the imperfect practical manifestation of these in actual relationships, and the possibility of alternatives, in which transaction and interest are also important elements.

The idealized picture of communal solidarity and mutual helpfulness is acted out only imperfectly in the real world – like everywhere else, people in rural Transkei have divergent and sometimes conflicting interests, certain households are avoided as un-cooperative or troublesome, co-operative arrangements break down, and there may be ill-feeling within sub-

---

4 Karp (1980, 83) has made a similar point in claiming that it is at beer drinks that Iteso 'represent to themselves contradictions in their social experience ..... [their] ideas about the nature of their social world and their experience of it'. Unfortunately Karp never demonstrates the link between beer drinking and economic practice. He shows that the rules of seating are based on the social identities of persons, and reconstruct these identities, but these are based on generation and kinship within the family, or (in the case of mortuary rituals) the wider kinship group. The sociability of beer drinking involves dropping everyday status distinctions, though this is not easy to achieve. People suspect that there may be underlying hostilities that are masked by the appearance of sociability and commensality. The rules of beer drinking are aimed at preventing violence and sorcery but these aspects of 'real life' are manifest at the beer party, which contains the potential for both. This negates the separation of beer drinking from the conflicts and animosities of everyday life.

wards and even sections. As indicated above, even these conflicts and disjunctures are brought into the structure of beer drinking where they are, at least temporarily, resolved (or publicly displayed) alongside the norm of interdependence and co-operation. In addition, a kind of balance is produced between territorial units through the reciprocal exchanges of beer, within which individual interests and power differences are submerged. And the possibility of reconfiguration is ever present. Beer for a work party also dramatizes the fragility of social relations and the imperfections of social actors, by publicizing changes in networks of inter-household economic co-operation. The following case illustrates some of these points, as well as making explicit the ideology of both neighbourhood and kinship, and the tolerance of imbalances (which is not infinite) that underlies some of the agricultural companies.

When Zimene (Cirha clan) brewed beer for those who ploughed his garden Ndoda (a Cirha clansman of Zimene's) was placed in charge of the event, as the head of the company that did the work. Zimene made this quite clear in his speeches, referring a number of times to his dependence on the goodwill of the Cirha clan, and referring to Ndoda as the 'prime minister' *(inkulumbuso)* of the beer drink. Ndoda replied that the ploughing group was grateful for the honour (*imbeko*) of being given charge of the beer. In fact, there were more non-Cirha men in Ndoda's company than Cirha, and perhaps Zimene was merely trying to stress the common clanship between himself and Ndoda (they were not of the same or demonstrably related lineages). In his initial address at the beer drink, Zimene referred to the fact that he had previously ploughed with Jozana, his next-door neighbour, of the Mkwayi clan. In very veiled terms he indicated that Jozana had not reciprocated as expected, staying at work for long periods while Zimene looked after Jozana's home and family. It had seemed to the other members of the Mkwayi clan that Jozana had absconded, and they gave up hope, but Zimene continued to act as a good neighbour. There came a point, however, when the imbalance could no longer be tolerated. Eventually, he said, he had decided to turn to the Cirha clan for help. A number of times he emphasized the point that he was a Cirha, and not a member of the Mkwayi clan. Now, as he put it, 'I do not lack [support from] the Cirha clan' (*Ndingadingi maCirha mna*).

As this case illustrates, the homestead is also defined in kinship terms, a point neglected in the analysis so far. The ideology of kinship is important to Shixini people, and it features, as we have seen, in the one form of co-operative work involving cattle. While of little consequence at beer drinking rituals, there are a variety of other rituals, of major importance in determining the relationship between living and dead, in which kin play a key role. Some of these are rituals, involving the killing of an ox or a goat, as well as the brewing of beer, directed at a particular deceased kinsman such as a former homestead head. They are aimed at securing his/her status as a communicating ancestor, and have little relevance for the nature of everyday social interaction. The major kinship rituals, those associated with the establishment of ancestorhood and the relationship between ancestors and their descendants, and those of a piacular nature, have a relatively atemporal and formal character. They stress genealogical order, the unity of the kin group and the relationship between living and dead, and have a timeless quality, largely unrelated to the actual process of everyday life, which is based on territorial relations and co-operation between neighbours. At these rituals, actual events in the past are referred to in a general, unspecified way and the same temporal quality is conveyed in references to 'custom' and 'tradition'. Practices such as the consecration of the ox, for example, are explained as being handed down from the remote past unchanged and unchanging. The effect is to fuse the timeless world of the ancestors with the world of the living. The ritual performance evokes the authority of the past and confers it on the present. The world of dreams and visions (often the motivating factor for ritual) is fused with the conscious world, and the ritual cycle is presented as fixed and timeless, a set sequence that must be followed generation after generation.[5]

Why should kinship remain important ideologically, and why should it continue to feature in relations of production even if it is only of minor

---

5 A detailed comparison of the major ancestor rituals with the rituals of beer drinking is contained in McAllister (1997).

importance in work parties? Partly because the relations of production *within* the household are kinship relations, and a great deal of everyday work gets done without labour co-operation between households. In addition, the homestead as kinship group is constituted in terms of exogamy rules that prohibit marriage within the clans of any of the four grandparents. Furthermore, inheritance is based on agnation, and agnatic kin are theoretically all potential heirs to each other's livestock and other wealth. So there are a variety of ways in which both agnation and other kinds of kinship remain important. In Shixini the person is constituted in terms of links defined as cognatic and affinal, as well as by links defined in terms of place. But land, a key resource, is no longer obtained through inheritance, but by virtue of being a member of a territorial community. Informally, a son frequently takes over the land allocated to his father once his parents have gone, but only if he remains locally resident is this possible. In this sense territory takes primacy over kinship in gaining access to land, as it does in working it.

In any case, culture often contains inconsistent or contradictory elements, and often provides alternatives, one or other of which may gain in importance as the context changes. In the Xhosa case, neighbourliness did not emerge as a *new* social principle with the decline in homestead size and the changing nature of rural production at the turn of the century, but it became more important and more prominent, and its reinforcement in beer drinks more frequent and more diverse, as the kinship system weakened and the mode of production changed.

Conclusion

It seems evident that beer-drinks provide a point at which everyday practice and ideas about society are brought together, a point at which human agency and social structure (mainly in the form of territorial organization) can be integrated and provided with supernatural sanction. But they also provide a point at which the potential for agency is renewed and redirected, to provide a pattern of action not necessarily identical to that that precedes it. But what are the more general lessons to be learnt from this analysis? What general conclusions can we come to about beer drinking, labour and agriculture? And what does the above analysis suggest for an understanding social life (and for 'development') in rural areas like the Transkei today?

Firstly, the Transkei ethnography confirms the growing tendency to view co-operative work in a multi-dimensional context that includes history, such as responses to ecological change and to colonialism, as the context in which work-parties emerge. To this we have to add political economy, or the wider institutional context, including, in this case, migrant labour, land alienation, segregationist policy, agricultural development, and so on, the context in which work parties persist, change or disappear. Finally, this takes place in accordance and in conjunction with local cultural characteristics, which may have to undergo re-emphasis and adaptation in response to the aforementioned. It is this that gives co-operative work its specific local character.

Secondly, the classification of types of co-operative work in terms of elements such as the strength of reciprocity and the presence or absence of festive elements may have outlived its usefulness, especially in view of the emphasis on placing different forms of co-operative work (in any one community) within a single analytical framework. In terms of our understanding of co-operative work in Africa I hope to have provided evidence that attention to the consumption (of beer) that accompanies this aspect of production (agricultural and other labour) is necessary for a proper understanding of the latter, even though co-operative labour takes on a variety of forms not all of which are associated with beer drinking. In the Transkei case, it also seems that one cannot make proper sense of beer-drinks in general, a significant form of consumption, without understanding how they are related to homestead production.

There has been an inordinate amount of attention paid to consumption and commodities in recent years in anthropology, and there is a need to bring production back into the picture. Richards (1939, 144) long ago made the tantalizing observation that among the Bemba '...the whole principle of the exchange of work for food is learnt as a practical expedient by children not yet in their teens, and it is on these reciprocal obligations to serve and to feed that the whole political and economic system of the tribe is based.' In other words, to interpret this in a contemporary way rather than literally, production and consumption are integrated in a holistic manner, and seem in some cases to provide the foundations of social and moral life.

It is worth considering the validity of such a proposition among Shixini people, at least at the local level, even if we have to reject it as being the basis for some larger 'social system'. It is evident in Shixini that consumption, of beer, maize, labour, and social capital, has to be placed in a wider framework which includes the of relations of production, in terms of which both material and social commodities are produced. The relationship between work and beer means that relations of production are embodied in the distribution of the commodities produced and in multiple acts of consumption. By linking the mobilization of labour to the consumption of the products of labour, one arrives a holistic picture of what makes for orderly social life in this community.

To express this slightly differently in the context of the field of material studies, the consumption of a particular commodity, beer, is a key element in the construction of human sociality. Beer is in this sense a social 'thing', as Appadurai put it, the exchange of which is used by people imaginatively to define themselves. Some have hailed consumption studies as a move away from a productionist bias and a concern with political economy. However, my analysis suggests that we cannot look only at the meanings generated by consumption patterns without relating this to social practice, in which production, or even political economy more generally, is crucial. Once we realize that the consumption of beer is part of a wider process of distribution and exchange which includes labour, it becomes possible to see the relationship between beer as commodity consumed and the relations through which its raw material, maize, are produced. This places labour and beer within a single local system.

In this connection, we have to return to the question lying dormant in chapter two, and to ask how it can be that a homestead mode of production characterized by a reliance on co-operative labour remains in evidence in Shixini and other parts of Transkei. After all, this is part of South Africa, the continent's most highly developed capitalist economy and also the country that stands out in terms of the length and intensity of oppression of the indigenous population. This takes us back to questions of the role of the state, the place of Transkei in the political economy of South Africa, and to questions of identity and resistance. Fortunately, all of these issues have been discussed in detail elsewhere, and need only be summarized here. It is well known that during the apartheid era and before, areas like Transkei were turned into 'labour reserves' from which to draw migrant workers to the mines and industries of 'white' South Africa. In addition, in the 1960s and 1970s, especially, parts of these areas were turned into 'dumping grounds' to which 'surplus' black people were relocated from other parts of the country. As indicated above, state support for agriculture and other forms of development in the Transkei has been negligible, and black rural areas in South Africa have been severely neglected in virtually every way. The people who lived in these areas were thus incorporated into the wider economy while being denied the benefits of citizenship and state support. Their reaction, as historians like William Beinart and anthropologists like Philip Mayer demonstrated, was to develop forms of resistance to incorporation, under such unfavourable terms, into the wider society, and to maintain homestead production and the rural social system as far as possible. Among conservative Xhosa, as mentioned above, this took the form of a traditionalist ideology known as 'redness' (*ubuqaba*), after the conservative custom of using clothes dyed with red ochre. 'Red' Xhosa were characterized by a fierce adherence to custom and a rejection of western ('white') ways. A vital part of this ideology was the devotion to rural Xhosa society and tradition, and especially to 'building the homestead', with all that this implies (Mayer 1980).

One of the major ways in which this response was threatened by the state was the introduction of agricultural 'betterment' schemes in the period 1950-1970, the effects of which were to disrupt forms of rural social organisation in black areas and to render homestead production even less viable than before. In the Transkei and elsewhere this was strongly resisted in the 1960s and 1970s, especially in the more conservative coastal districts like Gatyana. Even by the time apartheid collapsed and the Pretoria-controlled Transkei administration was disbanded, in the early 1990s, a large part of the Transkei (around 45%), concentrated in the coastal districts, had not been subjected to betterment (which by then had become known as 'planning and rehabilitation'). The Transkei administration had indeed developed plans for such schemes for most of the region not yet subjected to betterment, but the majority of these had not been implemented. As noted earlier, a plan developed for Shixini was partly implemented in the 1980s but was eventually abandoned, partly because of strong resistance from most of Shixini's people, who were well aware of the negative implications of the scheme for homestead production and economic well being.

The tenacity of the Red ideology even in situations where Transkei men spent a large part of their working lives in urban areas is remarkable — testimony both to their determination to maintain rural production and the unpalatability of apartheid. Philip Mayer and the other contributors to the 'Xhosa in Town' trilogy, published in the early 1960s, demonstrated how urban residence had by no means diminished the commitment to rural life on the part of Transkei migrants in East London. Mayer (1960) and Reader (1961) showed how red migrants remained oriented towards their country homes by maintaining a particular structure of social relationships that 'encapsulated' them in the city. They reproduced their rural social relations through forming links exclusively with their 'home boys' (*amakhaya*), with whom they lived in town. Within these closely knit social networks relationships were acted out and their ideological stance constructed primarily through communal cooking and eating among room-mates, and through membership of *iiseti*, drinking groups composed strictly of *amakhaya*. *Iiseti* were larger than room groups and embraced *amakhaya* from a wider rural neighbourhood. It was through involvement in this set of relationships, in which communal activity and generalised reciprocity prevailed, that the red migrant remained 'country-rooted', as Mayer put it.

Mayer predicted that the immanent destruction of the areas in which Red migrants lived and the creation of Mdantsane, the 'dormitory' city outside of East London, to which people were to be relocated in the 1960s in terms of the apartheid ideology, would make continued 'encapsulation' impossible in the future. However, Bank (1999) has recently demonstrated that the pattern of cultural reproduction of Red Xhosa conservativeness was maintained into the 1980s in East London. As Bank puts it, in this 'moral economy of collective trusteeship and joint responsibility ... sharing and a collective work ethic were ... embedded in the routines of everyday life, as was an ethos of generalised reciprocity ... The social glue which held these relations of trust together was a common commitment to building the *umzi*, providing unconditional support to *amakhaya*, and to respecting age as the key determinant of social standing ...' (*ibid.*, 409). Significantly for the purpose of our analysis of beer drinking and co-operative labour, this pattern of cultural reproduction was maintained, after relocation, through precisely the same two institutions as before, eating together and belonging to *iiseti*, though their relative importance changed.

In asking how the people of Shixini manage to maintain a largely agrarian lifestyle in the face of the momentous changes which have affected South Africa over the past thirty or forty years, the significance of maintaining a sense of community and rural Xhosa identity thus becomes clear, not simply as an ideal construct, but as one with practical implications. Identification with the land and its products and the great value placed on both cultivation and livestock husbandry is backed up with an ethic of mutual economic assistance, especially in the form of co-operative work, which makes agricultural production possible. A sense of community is maintained *despite and because of* the long-term and sometimes permanent emigration of many people. Since it became possible to urbanise legally in the 1980s, more and more Transkeians have been moving to the cities, and Shixini has been no exception to this. On the one hand such emigration threatens the social fabric because it removes people from the area, and it decreases the amount of skills and labour power available. On the other hand, it relieves the pressure on land that would otherwise inevitably result and, in the case of migrant workers who remit, helps provide for much needed cash income locally. Since land is available, the practice of agriculture is possible and, to many, profitable — not in the sense of generating money, but in terms of providing a food supply, though this is usually partial. Those who remain at home depend on each other in this respect, and the sense of community and social integrity is maintained through a dense network of relations which are based on neighbourhood and often reinforced by kinship, but which are manifested in economic practice. Not only does agriculture depend on community, therefore, but community also depends on agriculture. The two are in a mutually dependent and dialectical relationship with each other.

For Shixini people, their identity as Xhosa is closely bound up with their particular lifestyle, in which ties with the land are predominant. Cultivating the land is not simply a technical means of producing crops, it is also a cultural activity. Labouring in the fields or herding cattle are an opportunity for the expression of Xhosa values as well as economically important activities. As I hope I have demonstrated, these values are by no means static and unchanging. Identity is also asserted through contrasting rural agrarian activities with the urban 'other', i.e. life in town and the activities associated with wage earning. Even for those in town, the rural link and the importance of agrarian activity is manifested clearly in ritual. Firstly, rituals are ideally performed at ones rural home (of birth) and not in the city. Secondly, ritual involves the products of agricultural labour — maize (for beer) and livestock (for sacrifice). And brewing beer and offering animals to the ancestors at ones homestead is simultaneously a claim on and a legitimisation of certain rural rights and/or resources. It demonstrates that one 'belongs', not only to a specific kinship group but also to a specific territory of origin, and as such that one can legitimately claim the right to a homestead and arable land, as well as a right to the other natural resources of the area. To those who have left Shixini for a permanent life in town, this may not apply, but for those who remain, and who view their future as being linked to their Shixini homes, engagement in the urban world and earning money may well be absolute necessities, but they are not ultimate values in the sense that agriculture, animal husbandry and a homestead in Shixini are.

In the light of the above comments it seems appropriate to conclude with a final reference to the new agricultural and land reform policies being suggested by the South African state, discussed in chapter three against the background of the finding that agricultural production in Shixini, and probably in coastal Transkei in general, is considerably higher than was previously believed. The ecological adaptations that were made to facilitate the maintenance of household production at a reasonable level, particularly the switch from field cultivation to garden cultivation, were sketched in chapter two. In chapters five and six our concern has been with the social adaptations made by Shixini people to changing environmental, political, economic and demographic circumstances. What has become apparent here is that the maintenance of household production is embedded within a social and moral framework. This framework includes a territorial organisation and a system of land tenure that makes it possible for people to produce an agricultural output, despite the odds being stacked against them in terms of a complete lack of support from the state for their agrarian activities. It is one that has been developed under adverse circumstances, and which has persisted over a considerable period as a means to resist apartheid and construct a sense of meaning and identity, as well as to make optimal use of social and environmental resources to maintain homestead production. One has to ask whether the new land reform and agricultural development policy discussed in chapter three is likely to encourage and enhance agrarian activities among people like those living in Shixini. One fears that it will not. Instead, it seems calculated to undermine the social, moral and tenurial basis in terms of which rural society operates and agriculture is practised at the present.

References

ADAMS, M., B. COUSINS AND S. MANONA (2000), 'Land tenure and economic development in rural South Africa: Constraints and Opportunities.' In B. Cousins (Ed.) *At the Crossroads. Land and Agrarian Reform in South Africa into the 21st Century*, pp. 111-128. Cape Town: National Land Committee and Programme for Land and Agrarian Studies, University of the Western Cape.

AMES, D.W. (1958), 'Wolof co-operative work groups.' In W. Bascom and M. Herskovits (Eds.) *Continuity and Change in African Cultures*, pp. 224-237. Chicago: University of Chicago Press.

ANDREW, M. (1992), 'A geographical study of agricultural change since the 1930s in Shixini location, Gatyana district, Transkei.' Unpublished M.A. Thesis. Grahamstown: Rhodes University.

APPADURAI, A. (ed.). (1986), *The Social Life of Things: Commodities in Cultural Perspective*. Cambridge: Cambridge University Press.

ARDRI (1989), *Lima Development Report*. Alice: Agriculture and Rural development Research Institute, University of Fort Hare.

ASHTON, H. (1952), *The Basuto*. London: Oxford University Press.

BANK, L. (1999), 'Men with cookers: transformations in migrant culture, domesticity and identity in Duncan village, East London.' *Journal of Southern African Studies*, 25, 3, 393-416.

BARTH, F. (1967), 'Economic spheres in Darfur.' In R. Firth (Ed.) *Themes in Economic Anthropology*, pp. 149-174. London: Tavistock.

BAUMAN, R. AND C.L. BRIGGS (1990), 'Poetics and performance as critical perspectives on language and social life.' *Annual Review of Anthropology*, 19, 59-99.

BEINART W. (1982), *The Political Economy of Pondoland 1860 to 1930*. Johannesburg: Ravan Press.

BEINART, W. (1987), '*Amafelandawonye* (the die-hards): Popular protest and women's movements in Herschel district in the 1920s.' In W. Beinart and C. Bundy (Eds.) *Hidden struggles in Rural South Africa*, pp. 222-269. London: James Currey.

BEINART, W. (1992), 'Transkeian smallholders and agrarian reform.' *Journal of Contemporary African Studies*, 11, 2, 178-199.

BEMBRIDGE, T.J. (1990), 'Agricultural development in the developing areas of Southern Africa.' *Africa Insight*, 202, 1, 17-24.

BEMBRIDGE, T.J. (1991), 'Technology transfer in small-scale dryland crop production: Future challenges.' *Development Southern Africa*, 8, 4.

BERNSTEIN, H. (Ed.) (1996), *The Agrarian Question in South Africa*. London: Frank Cass.

BERNSTEIN, H. (1996a), 'The political economy of the maize *filitere*.' *Journal of Peasant Studies*, 23, 2/3, 120-133.

BERRY, S. (1993), *No Condition is Permanent: The Social Dynamics of Agrarian Change in Sub-Saharan Africa*. Madison: University of Wisconsin Press.

BLOCH, M. (1973), 'The long and the short term: the economic and political significance of the morality of kinship.' In J. Goody (Ed.) *The Character of Kinship*, pp. 75-87. Cambridge: Cambridge University Press.

BRYCESON, D. (1998), 'African rural labour: Where to?' Unpublished ms.

BURCHMORE, A. (1988), 'The soils and vegetation of Shixini administrative area: towards a land use plan.' B. A. (Hons.) dissertation. Rhodes University: Grahamstown.

CAIRNS, R.I. AND J.D. LEA (1990), 'An agricultural survey of subsistence farmers in the Nkandla district of KwaZulu.' *Development Southern Africa,* 7, 1.

CHARSLEY, S.R. (1976), 'The *silika*: a co-operative labour institution.' *Africa*, 46, 1, 34-47.

CLOETE, D. (1985), 'Maize production schemes in the Transkei.' Unpublished mimeo. Johannesburg.

COOK, P.A.W. (1931), *Social Organization and Ceremonial Institutions of the Bomvana*. Cape Town: Juta and Co.

COUSINS, B. (2000), 'Introduction: Does land and agrarian reform have a future and, if so, who will benefit?' In B. Cousins, (Ed.) *At the Crossroads. Land and Agrarian Reform in South Africa into the 21st Century,* pp. 1-8. Cape Town: National Land Committee and Programme for Land and Agrarian Studies, University of the Western Cape.

COUSINS, B. (2000a), 'Land reform at the crossroads: who will benefit?' *Reconstruct*, 20 February.

COUSINS, B. (2000b), Didiza's Recipe for Disaster. *Daily Mail and Guardian,* Internet Edition, 22 August.

CROSS, C. AND R. HAINES (Eds.) (1988), *Towards Freehold?* Cape Town: Juta and Co.

(THE) DAILY DISPATCH (2000), Internet Edition, 25/1/2000. See http://www.dispatch.co.za

(THE) DAILY DISPATCH (2000a), Internet Edition, 20/4/2000. See http://www.dispatch.co.za

DAVIES, C.S. (1927), 'Customs governing beer drinking among the amaBomvana.' *South African Journal of Science.* 24, 521-524.

DAVISON, J. (1993), 'Tenacious women: clinging to *banja* household production in the face of changing gender relations in Malawi.' *Journal of Southern African Studies*, 19, 3, 405-421.

DELIWE, D. (1997), 'The farmer support programme.' In C.J. De Wet and M.G. Whisson, *From Reserve to Region: Apartheid and Socil Change in the Keiskammahoek District of (former) Ciskei, 1950 to 1990*, pp. 69-95. Grahamstown: Institute of Social and Economic Research, Rhodes University. Occasional Paper No. 35.

DE WET, C.J. (1988), 'Land tenure, local government and agricultural development in the Ciskei.' In C. Cross and R. Haines (Eds.) *Towards Freehold?* pp. 348-356. Cape Town: Juta and Co.

DE WET, C.J. (1989), 'Betterment planning in a rural village in Keiskammahoek.' *Journal of Southern African Studies*, 15, 2, 326-345.

DOKE, C.M. (1931), *The Lambas of Northern Rhodesia*. London: George Harrap.

DOKE, C.M. AND B.W. VILAKAZI (1972), *Zulu-English Dictionary*. Second edition. Johannesburg: Witwatersrand University Press.

DONHAM, D.L. (1985, *Work and Power in Maale, Ethiopia*. Ann Arbor: UMI Research Press.

DOUGLAS, M. (1987), 'A distinct anthropological perspective.' In M. Douglas (Ed.) *Constructive Drinking: Perspectives on Drink from Anthropology*, pp. 3-15. Cambridge: Cambridge University Press.

ENGLUND , H. (1999), 'The self in self-interest: Land, labour and temporalities in Malawi's agrarian change.' *Africa*, 69, 1, 139-159.

ERASMUS, G.J. (1956), 'Culture, structure and process: The occurrence and disappearance of reciprocal farm labour.' *South Western Journal of Anthropology*, 12, 4, 444-469 .

FABIAN, J. (1990), *Power and Performance*. Madison: University of Wisconsin Press.

FAY, D. AND R. PALMER (2000), 'Prospects for redistribution of wealth through land reform in Dwesa-Cwebe.' In B. Cousins (Ed.) *At the Crossroads. Land and Agrarian Reform in South Africa into the 21st Century*, 194-210. Cape Town: National Land Committee and Programme for Land and Agrarian Studies, University of the Western Cape.

GARDINER, A. F. (1836), *Narrative of a Journey to the Zooloo Country in South Africa*. London: Crofts.

GESCHIERE, P. (1995), 'Working groups or wage labour? Cash-crops, reciprocity and money among the Maka of Southeastern Cameroon.' *Development and Change*, 26, 3, 503-523.

GOODFELLOW, D.M. (1939), *Principles of Economic Sociology*. London: George Routledge and Sons, Ltd.

GREGSON R.E. (1969), 'Beer, leadership and the efficiency of communal labour.' Unpublished conference paper. American Anthropological Association, 23 November.

GULLIVER, P. (1971), *Neigbours and Networks*. University of California Press.

HAMMOND-TOOKE, W.D. (1962), *Bhaca Society*. Cape Town: Oxford University Press.

HAMMOND-TOOKE, W.D. (1975), *Command or Consensus?* Cape Town: David Philip.

HAMMOND-TOOKE, W.D. (1984), 'In search of the lineage: the Cape Nguni case.' *Man* (n.s.), 19, 1, 77-93.

HERON, G. (1990), 'Household, production and the organization of co-operative labour in Shixini, Transkei.' Unpublished M. A. thesis. Rhodes University: Grahamstown.

HERON, G. AND L. CLOETE (1991), *Household Survival Strategies in Shixini, Transkei.* Pretoria: HSRC.

HOLY, L. (1977), 'Toka ploughing teams: towards a decision model of social recruitment.' In M. Stuchlik (Ed.) *Goals and Behaviour,* pp. 49-73. Belfast, Queens University, Department of Social Anthropology.

HUNTER, M. (1936), *Reaction to Conquest.* London: Oxford University Press.

ITURRA, R. (1977), 'Strategies in social recruitment: a case of mutual help in rural Galicia.' In M. Stuchlik (Ed.) *Goals and Behaviour.* pp. 75-94. Belfast: Queens University, Department of Social Anthropology.

JACKSON, A.O. (1975), *The Ethnic Composition of the Ciskei and Transkei.* Ethnological Publications No. 53. Pretoria: Government Printer.

JAMES, D. (1985), 'From co-operation to 'co-operative': changing patterns of agricultural work in a rural village.' Unpublished mimeo.

KARP, I. (1980), 'Beer Drinking and Social Experience in an African Society.' In I. Karp and C.S. Bird (Eds.) *Explorations in African Systems of Thought,* pp. 83-119. Indiana University Press.

KAY, S. (1833), *Travels and Researches in Caffraria.* London: John Mason.

KEPE, T. (2000), 'Clearing the ground in the spatial development initiatives (SDIs): analysing 'process' on South Africa's Wild Coast.' In B. Cousins (Ed.) *At the Crossroads. Land and Agrarian Reform in South Africa into the 21st Century,* pp. 254-263. Cape Town: National Land Committee and Programme for Land and Agrarian Studies, University of the Western Cape.

KINGON, J.R.L (1915), 'Native Agriculture'. In *Report of the South African Association for the Advancement of Science,* 12, 176-191.

KOPYTOFF, I. 'The Cultural Biography of Things: Commoditization as Process.' In A. Appadurai (Ed.) *The Social Life of Things: Commodities in Cultural Perspective,* pp. 64-91. Cambridge: Cambridge University Press.

KRIGE, E.J. AND J.D. KRIGE (1943), *The Realm of a Rain Queen.* London: Oxford University Press.

KROPF, A. (1915), *A Kafir-English Dictionary* (2nd Ed., edited by R Godfrey). Alice: Lovedale Press.

KUCKERTZ, H. (1983-4), 'Symbol and authority in Mpondo ancestor religion.' *African Studies,* 2, 42.

KUCKERTZ, H. (1985), 'Organizing labour forces in Pondoland: A new perspective on work-parties.' *Africa,* 55, 2, 115-131.

KUCKERTZ, H. (1990), *Creating Order: The Image of the Homestead in Mpondo Social Life.* Johannesburg: Witwatersrand University Press.

KUCKERTZ, H. (1997), '*Ukuhlonipha* as idiom of moral reasoning in Mpondo.' In P.A. McAllister (Ed.) *Culture and the Commonplace,* pp. 311-348. Johannesburg: Witwatersrand University Press.

KUPER, H. (1947), *The Swazi An African Aristocracy.* London: Oxford University Press.

LAHIFF, E. (2000), 'Response to the Ministry of Agriculture and Land Affairs' "Integrated Programme of Land Redistribution and Agricultural Development in South Africa" (8 June 2000)'. Nkuzi Development Association. 18 July.

LEEUWENBERG, J. (1977), *Transkei: A Study in Economic Regression*. London: The Africa Bureau.

LIEBENBERG, A. (1997), 'Dealing with relations of inequality: married women in a Transkei village.' In P.A. McAllister (Ed.) *Culture and the Commonplace*, pp. 349-374. Johannesburg: Witwatersrand University Press.

LIPTON, M. (1977), 'South Africa: Two agricultures?' In F. Wilson et al. (Eds.) *Farm Labour in South Africa*, pp. 72-85. Cape Town: David Philip.

LIPTON, M., F. ELLIS AND M. LIPTON (Eds.) (1996), *Land, Labour and Livelihoods in Rural South Africa*. Vol 2. Durban: Indicator Press.

MACHETE C.L. AND N.M. MOLLEL (2000), 'Extension and support services for smallholder agricultural development in South Africa: Who is the smallholder farmer?' In B. Cousins (Ed.) *At the Crossroads. Land and Agrarian Reform in South Africa into the 21st Century*, pp. 340-348. Cape Town: National Land Committee and Programme for Land and Agrarian Studies, University of the Western Cape.

MAHLATHI, V. (2000), 'Integrated planning and implementation in rural areas: experiences from agri-tourism spatial development initiatives (SDIs).' In B. Cousins (Ed.) *At the Crossroads. Land and Agrarian Reform in South Africa into the 21st Century*, pp. 248-253. Cape Town: National Land Committee and Programme for Land and Agrarian Studies, University of the Western Cape.

MARKS, S. AND A. ATMORE (1980), 'Introduction.' In S. Marks and A. Atmore (Eds.) *Economy and Society in Pre-Industrial South Africa*, pp. 1-43. London: Longmans.

MARWICK, M. (1966), *The Swazi*. London: Frank Cass.

MAY, J. (1996), 'Assets, incomes and livelihoods in rural KwaZulu-Natal.' In M. Lipton, F. Ellis and M. Lipton (Eds.). *Land, Labour and Livelihoods in Rural South Africa*. Vol 2. Durban: Indicator Press.

MAYER, P. (1951), *Two Studies in Applied Anthropology in Kenya*. London: HMSO.

MAYER, P. (1960), *Townsmen or Tribesmen?* Cape Town: Oxford University Press.

MAYER, P. (Ed.) (1980), *Black Villagers in an Industrial Society*. Cape Town: Oxford University Press.

MAYER, P. AND I. MAYER (1970), 'Socialisation by peers.' In P. Mayer (Ed.) *Socialisation, the Approach from Social Anthropology*, pp. 159-189. London: Tavistock.

MCALLISTER, P.A. (1979), 'The rituals of labour migration among the Gcaleka.' M.A. thesis, Grahamstown: Rhodes University.

MCALLISTER, P.A. (1980), 'Work, homestead and the shades: the ritual interpretation of migrant labour among the Gcaleka.' In P. Mayer (Ed.) *Black Villagers in an Industrial Society*, pp. 205-254. Cape Town: Oxford University Press.

MCALLISTER, P.A. (1981), *Umsindleko: A Gcaleka Ritual of Incorporation*. Grahamstown: Rhodes University, Institute of Social and Economic Research.

MCALLISTER, P.A. (1985), 'Beasts to beer pots: migrant labour and ritual change in Willowvale, Transkei.' *African Studies,* 44, 2, 1221-135.
MCALLISTER, P.A. (1986), 'Releasing the Widow — Xhosa Beer Drink Oratory and Status Change.' *African Studies,* 45, 2, 171-197.
MCALLISTER, P.A. (1987), 'Xhosa beer drinks and their oratory.' PhD thesis, Grahamstown, Rhodes University.
MCALLISTER, P.A. (1989), 'Resistance to 'Betterment' in the Transkei: A case study from Willowvale district.' *Journal of Southern African Studies,* 15, 2, 346-368.
MCALLISTER, P.A. (1992), 'Rural production, land use and development planning in Transkei: A critique of the Transkei Agricultural Development Study.' *Journal of Contemporary African Studies*, 11, 2, 200-222.
MCALLISTER, P.A. (1993), 'Indigenous beer in southern Africa: functions and fluctuations.' *African Studies,* 52, 1, 71-88.
MCALLISTER, P.A. (1993a), 'Public oratory in Xhosa ritual: tradition and change'. *Ethnology,* 32, 1, 291-304.
MCALLISTER, P.A. (1994), 'The Elwandle development research project: finding alternatives to 'betterment planning' in rural South Africa.' In D.J. Gouws (Ed.) *Case Studies in Research with a View to Implementation*, pp. 115-134. Pretoria: Human Sciences Research Council.
MCALLISTER, P.A. (1997), 'Ritual and social practice in the Transkei.' In P.A. McAllister (Ed.) *Culture and the Commonplace,* pp. 279-310. Johannesburg: Witwatersrand University Press.
MCALLISTER, P.A. AND D. DELIWE (1996), 'Youth in Rural Transkei: The Demise of 'Traditional' Youth Associations and the Development of New Forms of Association and Activity: 1975-1993.' Development Studies Working Paper No. 61, Grahamstown: Rhodes University, Institute of Social and Economic Research.
MCMILLAN, W. (1988), 'Small scale farming in Shixini, Willowvale district, Transkei.' Unpublished report. Grahamstown: Rhodes University.
MEAGHER, K. (1998), 'The developmental potential of the non-farm sector: Theoretical and policy issues.' Unpublished paper. Workshop on Deagrarianization and Rural Employment in Nigeria. Kano, 31 August — 2 September.
MOLL, T. (1984), 'A mixed and threadbare bag: Employment, incomes and poverty in Lower Roza, Qumbu, Transkei.' Cape Town: Carnegie Conference Paper No 47.
MOORE, M. P. (1975), 'Co-operative labour in peasant Agriculture.' *Journal of Peasant Studies,* 2, 3, 270-291.
MPAMBANI, S.A. (1994), *Poverty Profile of the Transkei Region*. Cape Town: University of Cape Town, SALDRU.
MURRAY, C. (1978), 'Migration, differentiation and the developmental cycle in Lesotho.' In W.M.J. van Binsbergen and H.A. Meilink (Eds.) *Migration and the Transformation of Modern African Society: African Perspectives*. Leiden: Afrika-Studiecentrum.
(NDALA) (1998), 'Agricultural Policy in South Africa – A Discussion Document'. Pretoria: Ministry for Agriculture and Land Affairs.

(NDALA) (2000), 'The Black Commercial Farmer Programme: A Component of an Integrated Land Redistribution Programme of the Ministry of Agriculture and Land Affairs.' Pretoria: National Departments of Agriculture and Land Affairs. 7 April.

(NDALA) (2000a), 'Integrated Programme of Land Redistribution and Agricultural Development in South Africa.' Pretoria: Ministry for Agriculture and Land Affairs.

(NKUZI) (2000), 'Response to the recent policy statement by the Minister for Agriculture and Land Affairs for strategic direction on land issues.' Nkuzi Development Association. 10 March.

OTTAWAY, M. (1996), 'Development without a peasantry: a forgotten aspect of the land use issue in South Africa.' *Development in Practice*, 6, 2, 129-139.

PAINTER,T.M. (1986), 'Spontaneous co-operation, introduced co-operation and agricultural development in south-western Niger.' In M.M. Horowitz and T.M. Painter (Eds.) *Anthropology and Rural Development in West Africa,* pp. 197-219. Boulder and London, Westview.

PALMER, R., H. TIMMERMANS AND D. FAY (Eds) (2000), *From Confrontation to Negotiation on South Africa's Wild Coast: Conservation, Land Reform and Tourism Development at Dwesa-Cwebe Nature Reserve.* Pretoria: Human Sciences Research Council. (Forthcoming).

PEIRES, J. (1981), *The House of Phalo.* Johannesburg: Ravan Press.

PHILLIPS-HOWARD, K.D. (1995), 'Small farm agriculture: Prospects and preconditions.' Umtata: Unpublished workshop paper.

PORTER, G. AND K.D. PHILLIPS-HOWARD (1997), 'Agricultural issues in the former homelands of South Africa: the Transkei.' *Review of African Political Economy*, 24, 72, 185-202.

READER, D.H. (1961), *The Black Man's Portion.* Cape Town: Oxford University Press.

RICHARDS, A. (1939), *Land, Labour and Diet in Northern Rhodesia.* London: Oxford University Press.

SANSOM, B. (1974), 'Traditional economic systems.' In W.D. Hammond-Tooke (Ed.) *The Bantu-Speaking Peoples of Southern Africa*, pp. 135-176. London: Routledge.

SAUL, M. (1983), 'Work parties, wages, and accumulation in a Voltaic village.' *American Ethnologist,* 10, 1, 77-96.

SHACKLETON, S. E., C.C. SHACKLETON AND B. COUSINS, B. (2000), 'The economic value of land and natural resources to rural livelihoods: Case studies from South Africa.' In B. Cousins (Ed.) *At the Crossroads. Land and Agrarian Reform in South Africa into the 21st Century,* pp. 35-67. Cape Town: National Land Committee and Programme for Land and Agrarian Studies, University of the Western Cape.

SHAW, M. (1974), 'Material Culture.' In W.D. Hammond-Tooke (Ed.) *The Bantu-Speaking Peoples of Southern Africa.* 'pp. 85-131. London: Routledge.

SHAW, M. AND VAN N.J. WARMELO (1981), 'The Material Culture of the Cape Nguni. Part 3, Subsistence.' *Annals of the South African Museum*, 58, 3, 215-445.

SIMON, C. (1989), 'Dealing with distress: a medical anthropological analysis of the search for health in a rural Transkeian village.' Unpublished Masters thesis. Grahamstown: Rhodes University.

SPIEGEL, A. (1980), 'Rural differentiation and the diffusion of migrant labour remittances in Lesotho.' In P. Mayer (Ed.) *Black Villagers in an Industrial Society*, pp. 109-169. Cape Town: Oxford University Press.

SWINDELL, K (1985), *Farm Labour*. Cambridge: Cambridge University Press.

TAMBIAH, S.J. (1981), 'A performative approach to ritual.' *Proceedings of the British Academy*. 1979, 65, 113-169.

(TRANSKEI) (1994), Population Census Report: Eastern Cape, Sub-Region Transkei. Vol 1, 1994. Transkei Central Statistics Office.

TURNER, V.W. (1957), *Schism and Continuity in an African Society*. Manchester: Manchester University Press.

UCHENDU, V.C. (1970), 'Traditional work groups in economic development.' In *Proceedings of the First Annual Conference of the Provisional Council for the Social Sciences in East Africa, University of Dar Es Salaam*, 248-261. Dar Es Salaam: Provisional Council for the Social Sciences in East Africa.

(UNIVERSITY OF NATAL) (1984), 'Annual report: Subsistence Agriculture Study Group.' University of Natal.

VAN WARMELO, N. J. (1974), 'The classification of cultural groups.' In W.D. Hammond-Tooke (Ed.) *The Bantu-Speaking Peoples of Southern Africa*, pp. 56-84. London: Routledge.

WEINRICH, A.K.H. (1975), *African Farmers in Rhodesia*. London: Oxford University Press.

WESTCOTT, G. (1977), 'Obstacles to agricultural development in the Transkei.' In F. Wilson et al. (Eds.) *Farm Labour in South Africa*, pp. 139-153. Cape Town: David Philip.

WHISKEN, J. (1995), 'An examination of the application of a geographical information system to rural development planning in Shixini location, Transkei.' Unpublished M. A. thesis. Grahamstown. Rhodes University.

WILSON, M. et al. (1952), *Social Structure: Keiskammahoek Rural Survey Vol 3*. Pietermaritzburg: Shuter and Shooter.

WILSON, M. (1971), 'The growth of peasant communities.' In M. Wilson and L. Thompson (Eds.) *The Oxford History of South Africa, vol 2,* pp. 49-103. London: Oxford University Press.

WORBY, E. (1995), 'What does agrarian wage labour signify? Cotton, commoditisation and social form in Gokwe, Zimbabwe.' *The Journal of Peasant Studies*. 23, 1, 1-129.

ASC Research Series
published by Ashgate Publishing Ltd.

1.	Dick Foeken & Nina Tellegen 1994	Tied to the land. Living conditions of labourers on large farms in Trans Nzoia District, Kenya
2.	Tom Kuhlman 1994	Asylum or aid? The economic integration of Ethiopian and Eritrean refugees in the Sudan
3.	Kees Schilder 1994	Quest for self-esteem. State, Islam and Mundang ethnicity in Northern Cameroon
4.	Johan A. van Dijk 1995	Taking the waters. Soil and water conservation among settling Beja nomads in Eastern Sudan
5.	Piet Konings, 1995	Gender and class in the tea estates of Cameroon
6.	Thera Rasing 1995	Passing on the rites of passage. Girls' initiation rites in the context of an urban Roman Catholic community on the Zambian Copperbelt
7.	Jan Hoorweg, Dick Foeken & Wijnand Klaver, 1995	Seasons and nutrition at the Kenya coast
8.	John A. Houtkamp 1996	Tropical Africa's emergence as a banana supplier in the inter-war period
9.	Victor Azarya 1996	Nomads and the state in Africa: the political roots of marginality
10.	Deborah Bryceson & Vali Jamal, eds., 1997	Farewell to farms. De-agrarianization and employment in Africa
11.	Tjalling Dijkstra 1997	Trading the fruits of the land: horticultural marketing channels in Kenya
12.	Nina Tellegen, 1997	Rural enterprises in Malawi: necessity or opportunity?
13.	Klaas van Walraven 1999	Dreams of power. The role of the Organization of African Unity in the politics of Africa, 1963-1993
14.	Isaac Sindiga 1999	Tourism and African development. Change and challenge of tourism in Kenya
15.	Laurens van der Laan, Tjalling Dijkstra & Aad van Tilburg, eds., 1999	Agricultural marketing in Tropical Africa. Contributions from the Netherlands

Copies can be ordered from:

Ashgate Publishing Direct Sales
Bookpoint Limited
39 Milton Park
Abingdon, Oxon, OX14 4TD
United Kingdom
Fax: +44 (0) 1235 400454
E-mail: orders@bookpoint.co.uk

For Product Safety Concerns and Information please contact our EU representative GPSR@taylorandfrancis.com Taylor & Francis Verlag GmbH, Kaufingerstraße 24, 80331 München, Germany

Batch number: 10704215

Printed by Printforce, the Netherlands